I0426934

Basic Physics: A Formula Handbook

N.B. Singh

DEDICATION

To Nature,

I dedicate this book to you, the source of all life. You are my inspiration, my teacher, and my friend.

Thank you for teaching me about the beauty of the world around me. Thank you for showing me the power of the natural world. Thank you for giving me a sense of peace and tranquillity.

I promise to do my part to protect you and your many wonders. I will teach my children about the importance of conservation and sustainability. I will work to make the world a better place for all living things.

Thank you for everything, Nature.

With love,

N.B Singh

Contents

11 Optics   105

12 Electromagnetic Waves   113

Preface

Welcome to "Basic Physics: A Formula Handbook"! This handbook is designed to serve as a comprehensive and concise reference for fundamental physics formulas and concepts. Whether you are a student, educator, or enthusiast, this book aims to provide a quick and accessible guide to essential physics knowledge.

Purpose of the Handbook

Physics is a subject that forms the foundation of many scientific disciplines. This handbook is created with the goal of offering a centralized resource for formulas commonly encountered in introductory and intermediate physics courses. It covers a broad range of topics, including mechanics, electromagnetism, thermodynamics, optics, and modern physics.

Organization

The handbook is organized by chapters, each focusing on a specific branch of physics. Within each chapter, you will find formulas, explanations, and examples to help you understand and apply the principles covered. The layout is designed to facilitate easy navigation and quick reference.

How to Use This Handbook

Whether you are studying for an exam, working on a physics problem, or just refreshing your knowledge, this handbook is here to assist you. Each formula is presented with its derivation, where applicable, and practical examples to illustrate its application. The goal is to enhance your understanding of physics and foster a deeper appreciation for the subject.

Acknowledgments

Thank you for choosing "Basic Physics: A Formula Handbook." May your journey through the fascinating world of physics be both enriching and enjoyable!

Chapter 1

Introduction to Physics

1.1 Overview of Physics

Physics is a fundamental branch of science that seeks to understand the fundamental principles governing the universe. It encompasses a wide range of phenomena, from the smallest subatomic particles to the vast reaches of the cosmos. In this section, we will provide a brief overview of the key concepts in physics.

1.1.1 Fundamental Concepts

Physics is built upon a few fundamental concepts that form the basis for understanding the natural world. These include:

1. Space and Time

Einstein's theory of relativity revolutionized our understanding of space and time. The famous equation $E = mc^2$ relates energy to mass and the speed of light, highlighting the interplay between these fundamental entities.

2. Forces and Motion

Newton's laws of motion describe the relationship between the motion of an object and the forces acting upon it. The second law, $F = ma$, quantifies the effect of force on an object's acceleration.

1.1.2   Example: Newton's Second Law

Consider a mass m subjected to a force F. According to Newton's second law, the acceleration (a) of the mass is given by the formula $F = ma$. For instance, if a force of 10 N is applied to a mass of 2 kg, the resulting acceleration can be calculated as:

$$a = \frac{F}{m} = \frac{10\,\text{N}}{2\,\text{kg}} = 5\,\text{m/s}^2$$

This example illustrates the relationship between force, mass, and acceleration.

1.1.3   3. Energy and Matter

Einstein's equation $E = mc^2$ demonstrates the equivalence of energy (E) and mass (m). It reveals that a small amount of mass can be converted into a large amount of energy, providing insights into nuclear reactions and the energy released in processes such as nuclear fission.

1.2   Scientific Method

The scientific method is a systematic approach used by scientists to investigate natural phenomena, acquire knowledge, and develop theories. It involves a series of steps designed to ensure objectivity, reproducibility, and reliability in the pursuit of scientific understanding.

1.2.1 Steps of the Scientific Method

The scientific method typically involves the following steps:

1. Observation

Scientific inquiry begins with careful observation of a phenomenon or event. Observations lead to questions about how and why things happen.

2. Formulation of a Hypothesis

A hypothesis is a testable explanation for the observed phenomenon. It is a proposed solution that can be evaluated through experimentation.

3. Experimental Design

Experiments are designed to test the hypothesis. This includes defining variables, controlling conditions, and collecting data.

4. Data Collection

Experimental data is collected systematically using appropriate tools and techniques. Precise measurements are crucial for the reliability of results.

5. Analysis and Interpretation

The collected data is analyzed using statistical methods and interpreted to draw conclusions. This step involves identifying patterns and relationships.

6. Conclusion and Communication

Conclusions are drawn based on the analysis. The findings are communicated through scientific reports, papers, or presentations.

1.2.2   Example: Experimental Design

Consider an experiment to investigate the effect of varying the angle of launch on the range of a projectile. The following variables are identified:

- Independent Variable: Angle of Launch (θ) - Dependent Variable: Range (R) - Control Variable: Initial Velocity (v_0)

The hypothesis could be that the range is maximized at a specific launch angle. The experiment involves launching projectiles at different angles and measuring the corresponding ranges.

The range (R) can be calculated using the projectile motion formula:

$$R = \frac{v_0^2 \sin(2\theta)}{g}$$

Where: v_0 is the initial velocity, θ is the launch angle, and g is the acceleration due to gravity.

1.2.3   Data Collection and Analysis

Suppose the experiment is conducted, and the following data is collected:

Launch Angle (θ)	Range (R)
30°	15 m
45°	20 m
60°	18 m

The data can be analyzed to determine the launch angle that maximizes the range. In this example, it is observed that the range is highest at a 45° launch angle.

1.3   Units and Measurement

Units and measurements are fundamental aspects of physics, providing the foundation for expressing quantities and understanding the physical world. In this section, we explore the concepts of units, dimensions, and measurement systems.

1.3.1 Fundamental Units

The International System of Units (SI) defines seven base units, from which all other units are derived:

- Meter (m) - for length

- Kilogram (kg) - for mass

- Second (s) - for time

- Ampere (A) - for electric current

- Kelvin (K) - for temperature

- Mole (mol) - for amount of substance

- Candela (cd) - for luminous intensity

These base units form the building blocks for measuring various physical quantities.

1.3.2 Derived Units

Derived units are combinations of base units. For example, the unit of speed is meters per second (m/s), which is derived from the base units of length and time.

1.3.3 Dimensional Analysis

Dimensional analysis is a powerful tool in physics. It involves examining the dimensions (units) of physical quantities to check the consistency of equations and derive relationships. For example, the equation for velocity (v) is given by:

$$v = \frac{\text{distance}}{\text{time}}$$

Using dimensional analysis, we confirm that the units of velocity are indeed $\frac{\text{meters}}{\text{seconds}}$.

1.3.4   Example: Speed Calculation

Consider an object moving with a constant speed of 20 m/s. If it travels for 30 seconds, the distance covered (d) can be calculated using the formula:

$$d = v \cdot t$$

where: v is the speed (20 m/s) and t is the time (30 s).

Substituting the values:

$$d = (20 \, \text{m/s}) \cdot (30 \, \text{s}) = 600 \, \text{meters}$$

So, the object covers a distance of 600 meters in 30 seconds.

1.3.5   Measurement Systems

Different measurement systems exist, and conversions between them are essential. The most commonly used systems include the metric system (SI), imperial system, and US customary system. Converting between units requires knowledge of conversion factors.

1.3.6   Example: Unit Conversion

To convert a length from meters to kilometers, the conversion factor is 1 kilometer = 1000 meters. If a distance is 5000 meters, the equivalent distance in kilometers (d_{km}) is given by:

$$d_{\text{km}} = \frac{5000 \, \text{m}}{1000 \, \text{m/km}} = 5 \, \text{km}$$

1.3.7   Precision and Accuracy

Precision refers to the reproducibility of measurements, while accuracy is the closeness of a measurement to its true value. The significant figures in a measurement indicate its precision. For instance, a length measured as 25.0 meters implies greater precision than 25 meters.

1.3.8 Example: Precision and Accuracy

If a target has a true diameter of 10.0 cm, and three measurements are made with calipers resulting in 9.8 cm, 9.9 cm, and 10.2 cm, the measurements are precise (consistent with each other) but not accurate (not close to the true value).

1.3.9 Errors in Measurement

Errors in measurement can be systematic or random. Systematic errors result from flaws in the measurement system, while random errors are unpredictable variations. Techniques such as averaging multiple measurements can help mitigate random errors.

1.3.10 Example: Systematic Error

If a scale consistently reads 2 grams heavier than the true weight, every measurement on that scale will have a systematic error of +2 grams.

1.3.11 Significant Figures

Significant figures indicate the precision of a measured quantity. Rules for determining significant figures include: - Non-zero digits are always significant. - Any zeros between significant digits are significant. - Leading zeros (zeros to the left of the first non-zero digit) are not significant. - Trailing zeros (zeros to the right of the last non-zero digit) in a decimal number are significant.

1.3.12 Example: Determining Significant Figures

For the number 0.00345, there are three significant figures.

1.3.13 Error Analysis

Error analysis involves determining the uncertainty in a measurement. The percent error (%error) is calculated as:

$$\%\text{error} = \left| \frac{\text{experimental value} - \text{accepted value}}{\text{accepted value}} \right| \times 100$$

1.3.14   Example: Percent Error Calculation

If the accepted value of a measurement is 50 cm, and the experimental measurement is 48 cm, the percent error is:

$$\%\text{error} = \left| \frac{48\,\text{cm} - 50\,\text{cm}}{50\,\text{cm}} \right| \times 100$$

1.3.15   Uncertainty and Significant Figures

The last digit in a measurement is uncertain, and the uncertainty is often expressed using the term "significant figures." For instance, if a length is measured as 25.0 cm, the last digit "0" indicates the uncertainty, and the measurement is considered to have three significant figures.

1.3.16   Example: Expressing Uncertainty

If the temperature is measured as 23.5°C, the uncertainty in the measurement is ±0.1°C. Therefore, the result can be expressed as 23.5 ± 0.1°C.

1.4   Vectors in Physics

Vectors are fundamental mathematical entities in physics, representing quantities that have both magnitude and direction. Understanding vector concepts is crucial for describing motion, forces, and various physical phenomena.

1.4.1   Vector Representation

A vector is typically represented by an arrow in a specific direction, where the length of the arrow corresponds to the magnitude of the vector, and the arrow's direction indicates the vector's direction. In Cartesian coordinate systems, vectors are often expressed in terms of their components.

1.4.2 Vector Notation

In vector notation, a vector is denoted by a boldface letter or an arrow above the letter. For example, $\mathbf{A}$ or $\vec{B}$ represents a vector quantity.

1.4.3 Vector Addition

Vector addition is performed by adding corresponding components. For two vectors $\mathbf{A} = (A_x, A_y, A_z)$ and $\mathbf{B} = (B_x, B_y, B_z)$, their sum $\mathbf{C} = \mathbf{A} + \mathbf{B}$ is given by:

$$C_x = A_x + B_x, \quad C_y = A_y + B_y, \quad C_z = A_z + B_z$$

1.4.4 Example: Vector Addition

Consider two vectors $\mathbf{A} = (3, -2, 1)$ and $\mathbf{B} = (-1, 4, 2)$. The sum $\mathbf{C} = \mathbf{A} + \mathbf{B}$ is calculated as:

$$C_x = 3 + (-1) = 2, \quad C_y = (-2) + 4 = 2, \quad C_z = 1 + 2 = 3$$

So, $\mathbf{C} = (2, 2, 3)$.

1.4.5 Scalar Multiplication

Scalar multiplication involves multiplying a vector by a scalar. For a vector $\mathbf{A} = (A_x, A_y, A_z)$ and a scalar k, the scalar multiple $k\mathbf{A}$ is given by:

$$k\mathbf{A} = (kA_x, kA_y, kA_z)$$

1.4.6 Example: Scalar Multiplication

If $\mathbf{A} = (2, -1, 3)$ and $k = 3$, the scalar multiple $k\mathbf{A}$ is:

$$k\mathbf{A} = 3 \cdot (2, -1, 3) = (6, -3, 9)$$

1.4.7   Dot Product

The dot product (or scalar product) of two vectors $\mathbf{A}$ and $\mathbf{B}$ is defined as:

$$\mathbf{A} \cdot \mathbf{B} = A_x B_x + A_y B_y + A_z B_z$$

1.4.8   Example: Dot Product

For vectors $\mathbf{A} = (1, -2, 3)$ and $\mathbf{B} = (4, 0, -2)$, the dot product $\mathbf{A} \cdot \mathbf{B}$ is calculated as:

$$\mathbf{A} \cdot \mathbf{B} = (1)(4) + (-2)(0) + (3)(-2) = 1 - 6 = -5$$

1.4.9   Cross Product

The cross product (or vector product) of two vectors $\mathbf{A}$ and $\mathbf{B}$ is given by a vector $\mathbf{C} = \mathbf{A} \times \mathbf{B}$ with components:

$$C_x = A_y B_z - A_z B_y, \quad C_y = A_z B_x - A_x B_z, \quad C_z = A_x B_y - A_y B_x$$

1.4.10   Example: Cross Product

For vectors $\mathbf{A} = (2, 3, -1)$ and $\mathbf{B} = (4, -1, 2)$, the cross product $\mathbf{C} = \mathbf{A} \times \mathbf{B}$ is calculated as:

$$C_x = (3)(2) - (-1)(-1) = 5, \quad C_y = (-1)(4) - (2)(2) = -8,$$
$$C_z = (2)(-1) - (3)(4) = -14$$

So, $\mathbf{C} = (5, -8, -14)$.

1.4.11   Unit Vectors

Unit vectors are vectors with a magnitude of 1. The unit vectors in the Cartesian coordinate system are denoted as $\mathbf{i}$, $\mathbf{j}$, and $\mathbf{k}$ for the x, y, and z directions, respectively.

1.4.12 Example: Unit Vector

The unit vector in the direction of the vector $\mathbf{D} = (3, 4, -2)$ is calculated as:

$$\mathbf{u} = \frac{\mathbf{D}}{|\mathbf{D}|} = \frac{(3, 4, -2)}{\sqrt{3^2 + 4^2 + (-2)^2}} = \left(\frac{3}{7}, \frac{4}{7}, -\frac{2}{7}\right)$$

1.4.13 Vector Components

The components of a vector $\mathbf{E}$ along the x, y, and z axes are denoted as E_x, E_y, and E_z, respectively. These components are determined by projecting the vector onto each axis.

1.4.14 Example: Vector Components

For a vector $\mathbf{F} = (4, -3, 5)$, its components along the x, y, and z axes are:

$$F_x = 4, \quad F_y = -3, \quad F_z = 5$$

1.4.15 Vector Magnitude

The magnitude (or length) of a vector $\mathbf{G} = (G_x, G_y, G_z)$ is given by:

$$|\mathbf{G}| = \sqrt{G_x^2 + G_y^2 + G_z^2}$$

1.4.16 Example: Vector Magnitude

For a vector $\mathbf{H} = (3, -1, 2)$, its magnitude is calculated as:

$$|\mathbf{H}| = \sqrt{3^2 + (-1)^2 + 2^2} = \sqrt{14}$$

1.4.17 Vector Projection

The projection of a vector $\mathbf{I}$ onto another vector $\mathbf{J}$ is given by:

$$\text{proj}_{\mathbf{J}}(\mathbf{I}) = \frac{\mathbf{I} \cdot \mathbf{J}}{|\mathbf{J}|} \cdot \frac{\mathbf{J}}{|\mathbf{J}|}$$

1.4.18   Example: Vector Projection

If $\mathbf{K} = (2, 1, -3)$ is projected onto $\mathbf{L} = (-1, 2, 4)$, the projection is calculated as:

$$\text{proj}_{\mathbf{L}}(\mathbf{K}) = \frac{(2, 1, -3) \cdot (-1, 2, 4)}{\sqrt{(-1)^2 + 2^2 + 4^2}} \cdot \frac{(-1, 2, 4)}{\sqrt{(-1)^2 + 2^2 + 4^2}}$$

1.4.19   Vector Angle

The angle θ between two vectors $\mathbf{M}$ and $\mathbf{N}$ is given by the arccosine of the dot product divided by the product of their magnitudes:

$$\cos(\theta) = \frac{\mathbf{M} \cdot \mathbf{N}}{|\mathbf{M}| \cdot |\mathbf{N}|}$$

1.4.20   Example: Vector Angle

For vectors $\mathbf{P} = (1, -1, 2)$ and $\mathbf{Q} = (3, 0, 1)$, the angle θ between them is calculated as:

$$\cos(\theta) = \frac{(1)(3) + (-1)(0) + (2)(1)}{\sqrt{1^2 + (-1)^2 + 2^2} \cdot \sqrt{3^2 + 0^2 + 1^2}}$$

1.4.21   Vector Cross Product and Area

The magnitude of the cross product $\mathbf{R} = \mathbf{S} \times \mathbf{T}$ represents the area of the parallelogram formed by vectors $\mathbf{S}$ and $\mathbf{T}$.

1.4.22   Example: Cross Product and Area

For vectors $\mathbf{U} = (2, -1, 3)$ and $\mathbf{V} = (4, -1, 2)$, the magnitude of their cross product $\mathbf{W} = \mathbf{U} \times \mathbf{V}$ gives the area of the parallelogram:

$$|\mathbf{W}| = |\mathbf{U} \times \mathbf{V}| = \sqrt{5^2 + (-8)^2 + (-14)^2}$$

1.4.23 Vector Applications

Vectors find applications in various physics topics, including kinematics, dynamics, electromagnetism, and fluid dynamics. They provide a concise and powerful way to represent physical quantities with both magnitude and direction.

1.4.24 Example: Velocity Vector

In kinematics, the velocity of an object is represented by a vector $\mathbf{V}$ with components V_x, V_y, and V_z representing the object's speed in each dimension.

1.5 Error Analysis and Measurement Uncertainty

Error analysis is an essential aspect of experimental physics, helping to quantify and understand the uncertainties associated with measurements. Accurate measurements are crucial in obtaining reliable results and drawing meaningful conclusions in scientific experiments.

1.5.1 Types of Errors

Errors in measurements can be categorized into two main types: systematic errors and random errors.

Systematic Errors

Systematic errors are consistent and repeatable errors that affect measurements in the same way each time. These errors can result from equipment limitations, environmental conditions, or procedural flaws. Systematic errors can often be identified and corrected.

Random Errors

Random errors, also known as statistical errors, are unpredictable variations in measurements. They can result from factors such as instrument precision, environmental noise, or fluctuations in experimental conditions. Random errors are typically analyzed using statistical methods.

1.5.2 Absolute and Relative Errors

Absolute Error

The absolute error (E) of a measurement is the magnitude of the difference between the measured value (X_{measured}) and the true value (X_{true}):

$$E = |X_{\text{measured}} - X_{\text{true}}|$$

Relative Error

The relative error (RE) is the ratio of the absolute error to the true value, often expressed as a percentage:

$$RE = \left| \frac{X_{\text{measured}} - X_{\text{true}}}{X_{\text{true}}} \right| \times 100\%$$

1.5.3 Example: Absolute and Relative Error

If a measured length is 20.5 cm, and the true length is 20.0 cm, the absolute error is $|20.5 - 20.0| = 0.5$ cm. The relative error is $\left| \frac{20.5 - 20.0}{20.0} \right| \times 100\% = 2.5\%$.

1.5.4 Precision and Accuracy

Precision

Precision refers to the reproducibility and consistency of measurements. High precision implies small random errors and tight clustering of measured values.

Accuracy

Accuracy is the closeness of measurements to the true or accepted value. Accurate measurements have small systematic errors.

1.5.5 Example: Precision and Accuracy

Consider a series of weight measurements with a balance. If the measurements consistently cluster around the true value, the results are precise. If the average of the measurements is close to the true weight, the results are accurate.

1.5.6 Standard Deviation and Variance

Standard Deviation

The standard deviation (σ) measures the spread or dispersion of a set of measurements. It is calculated using the formula:

$$\sigma = \sqrt{\frac{\sum_{i=1}^{N}(X_i - \bar{X})^2}{N}}$$

where N is the number of measurements, X_i is each individual measurement, and $\bar{X}$ is the mean.

Variance

The variance (σ^2) is the square of the standard deviation.

$$\sigma^2 = \frac{\sum_{i=1}^{N}(X_i - \bar{X})^2}{N}$$

1.5.7 Example: Standard Deviation and Variance

For a set of measurements (3.2, 3.5, 3.3, 3.4), the mean ($\bar{X}$) is 3.35. The standard deviation is calculated using the formula.

1.5.8 Propagation of Errors

When quantities are combined in calculations, the errors associated with each quantity contribute to the overall uncertainty in the result. The propagation of errors is often analyzed using calculus.

1.5.9   Example: Propagation of Errors

If the area (A) of a rectangle is calculated as $A = L \times W$, and the lengths have uncertainties (dL and dW), the uncertainty in the area (dA) can be determined using calculus.

1.5.10   Significant Figures and Error Reporting

Significant Figures in Calculations

When performing calculations with measured values, the result should be reported with the same number of significant figures as the least precise measurement.

Error Reporting

Errors should be reported to the same decimal place as the least precise measurement.

1.5.11   Example: Significant Figures and Error Reporting

If the length of a rod is measured as 12.345 cm with an uncertainty of ± 0.01 cm, the result of a calculation involving this length should be reported with the same number of decimal places, and the uncertainty should be ± 0.01 cm.

1.5.12   Calibration and Error Reduction

Regular calibration of instruments is crucial to minimize systematic errors. Calibration involves comparing instrument readings to known standards.

1.5.13   Example: Calibration

If a thermometer is calibrated using a standard thermometer, any systematic error can be identified and corrected.

1.5.14 Error Bars in Graphs

In graphical representation, error bars are often used to visually convey the uncertainty associated with data points.

1.5.15 Example: Error Bars

A scatter plot with error bars allows the viewer to assess the range of uncertainty associated with each data point.

Chapter 2

Kinematics

2.1 Motion in One Dimension

Motion in one dimension refers to the movement of an object along a straight line. The fundamental parameters describing such motion include position, displacement, velocity, and acceleration.

2.1.1 Position and Displacement

Position (x)

Position represents the location of an object in space at a particular time. It is usually measured with respect to a reference point. The SI unit of position is meters (m).

Displacement (Δx)

Displacement is the change in position and is a vector quantity. It is the difference between the final position (x_f) and the initial position (x_i).

$$\Delta x = x_f - x_i$$

Average Velocity (v_{avg})

Average velocity is the ratio of displacement to the time interval over which the displacement occurs.

$$v_{\text{avg}} = \frac{\Delta x}{\Delta t}$$

where Δt is the time interval.

2.1.2   Example: Average Velocity

If an object moves from $x_i = 2\,\text{m}$ to $x_f = 8\,\text{m}$ in $t = 4\,\text{s}$, the average velocity is:

$$v_{\text{avg}} = \frac{8\,\text{m} - 2\,\text{m}}{4\,\text{s}} = \frac{6\,\text{m}}{4\,\text{s}} = 1.5\,\text{m/s}$$

2.1.3   Instantaneous Velocity (v)

Instantaneous velocity is the velocity of an object at a specific moment in time. It is the limiting case of average velocity as the time interval approaches zero.

$$v = \lim_{\Delta t \to 0} \frac{\Delta x}{\Delta t} = \frac{dx}{dt}$$

2.1.4   Acceleration (a)

Acceleration is the rate of change of velocity with respect to time. It is a vector quantity. The SI unit of acceleration is meters per second squared (m/s^2).

$$a = \frac{\Delta v}{\Delta t}$$

2.1.5   Example: Acceleration

If the velocity of an object changes from $v_i = 2\,\text{m/s}$ to $v_f = 6\,\text{m/s}$ in $t = 3\,\text{s}$, the acceleration is:

$$a = \frac{6\,\text{m/s} - 2\,\text{m/s}}{3\,\text{s}} = \frac{4\,\text{m/s}}{3\,\text{s}}$$

2.1.6 Equations of Motion

For uniformly accelerated motion, the following equations relate displacement (Δx), initial velocity (v_i), final velocity (v_f), acceleration (a), and time (t):

First Equation of Motion

$$v_f = v_i + at$$

Second Equation of Motion

$$\Delta x = v_i t + \frac{1}{2}at^2$$

Third Equation of Motion

$$v_f^2 = v_i^2 + 2a\Delta x$$

2.1.7 Example: Equations of Motion

Suppose an object accelerates from rest ($v_i = 0$) with $a = 2\,\mathrm{m/s^2}$ for $t = 3\,\mathrm{s}$. Using the first equation of motion, the final velocity is:

$$v_f = 0 + (2\,\mathrm{m/s^2})(3\,\mathrm{s}) = 6\,\mathrm{m/s}$$

Using the second equation of motion, the displacement is:

$$\Delta x = 0 + \frac{1}{2}(2\,\mathrm{m/s^2})(3\,\mathrm{s})^2 = 9\,\mathrm{m}$$

2.1.8 Free Fall and Gravitational Acceleration

In free fall near the Earth's surface, objects experience constant acceleration due to gravity (g). The value of g is approximately $9.8\,\mathrm{m/s^2}$ downward.

$$g = 9.8\,\mathrm{m/s^2}$$

2.1.9   Example: Free Fall

If an object is dropped from a height of $h = 50\,\text{m}$, the time taken to reach the ground can be calculated using the second equation of motion:

$$h = \frac{1}{2}gt^2$$

Solving for t:

$$t = \sqrt{\frac{2h}{g}} = \sqrt{\frac{2 \times 50\,\text{m}}{9.8\,\text{m/s}^2}}$$

2.1.10   Projectile Motion

Projectile motion involves the motion of an object launched into the air, experiencing both horizontal and vertical motion. The horizontal motion is uniform, while the vertical motion is uniformly accelerated due to gravity.

2.1.11   Example: Projectile Motion

Consider a projectile launched with an initial velocity (v_i) at an angle (θ) above the horizontal. The horizontal and vertical components of velocity are given by:

$$v_{i_x} = v_i \cos(\theta)$$

$$v_{i_y} = v_i \sin(\theta)$$

The time of flight (T), horizontal range (R), and maximum height (H) can be calculated using kinematic equations.

2.1.12   Relative Motion

Relative motion deals with the motion of one object with respect to another. The relative velocity (v_{rel}) of object B with respect to object A is given by:

$$v_{\text{rel}} = v_{\text{B}} - v_{\text{A}}$$

2.1.13   Example: Relative Motion

If two cars A and B are moving with velocities $v_A = 20\,\text{m/s}$ and $v_B = 15\,\text{m/s}$ in the same direction, the relative velocity of B with respect to A is:

$$v_{\text{rel}} = v_B - v_A = 15\,\text{m/s} - 20\,\text{m/s} = -5\,\text{m/s}$$

2.2   Motion in Two Dimensions

Motion in two dimensions involves the simultaneous motion of an object along both the x and y axes. Key concepts include vectors, vector addition, projectile motion, and the resolution of vectors into components.

2.2.1   Vector Representation

Vectors are quantities with both magnitude and direction. In two-dimensional motion, position, velocity, and acceleration are represented as vectors. The general form of a vector $\mathbf{V}$ is:

$$\mathbf{V} = V_x \hat{i} + V_y \hat{j}$$

where V_x and V_y are the components along the x and y axes, and $\hat{i}$ and $\hat{j}$ are unit vectors in the x and y directions, respectively.

2.2.2   Vector Addition

Vector addition involves combining vectors to find their resultant. The resultant vector $\mathbf{R}$ of two vectors $\mathbf{A}$ and $\mathbf{B}$ is given by:

$$\mathbf{R} = \mathbf{A} + \mathbf{B}$$

The components of the resultant vector are the sum of the corresponding components of the individual vectors.

2.2.3   Example: Vector Addition

If $\mathbf{A} = 3\hat{i} + 4\hat{j}$ and $\mathbf{B} = -2\hat{i} + 1\hat{j}$, the resultant vector $\mathbf{R} = \mathbf{A} + \mathbf{B}$ is:

$$\mathbf{R} = (3 - 2)\hat{i} + (4 + 1)\hat{j} = \hat{i} + 5\hat{j}$$

2.2.4   Projectile Motion

Projectile motion occurs when an object is launched into the air and follows a curved path under the influence of gravity. The motion can be analyzed independently along the x and y axes.

Horizontal Motion

The horizontal motion is uniform and can be described by the equation:

$$x = x_0 + v_{0x}t$$

where x_0 is the initial x-coordinate, v_{0x} is the initial x-component of velocity, and t is time.

Vertical Motion

The vertical motion is uniformly accelerated due to gravity. The equations of motion are:

$$y = y_0 + v_{0y}t - \frac{1}{2}gt^2$$

$$v_y = v_{0y} - gt$$

where y_0 is the initial y-coordinate, v_{0y} is the initial y-component of velocity, g is the acceleration due to gravity, and v_y is the y-component of velocity.

2.2.5   Example: Projectile Motion

Consider a projectile launched with an initial speed of $v_0 = 20\,\text{m/s}$ at an angle of $30°$ above the horizontal. The initial components of velocity are:

$$v_{0x} = v_0 \cos(30°)$$

$$v_{0y} = v_0 \sin(30°)$$

The time of flight (T), horizontal range (R), and maximum height (H) can be calculated using kinematic equations.

2.2.6   Resolution of Vectors

The resolution of vectors involves determining their components along specified axes. The components can be found using trigonometric functions.

Horizontal Component (V_x)

$$V_x = V \cos(\theta)$$

Vertical Component (V_y):

$$V_y = V \sin(\theta)$$

where V is the magnitude of the vector, and θ is the angle of the vector with the positive x-axis.

2.2.7   Example: Resolution of Vectors

If a velocity vector $\mathbf{V} = 15\hat{i} + 10\hat{j}$, the magnitude of the velocity and its angle with the positive x-axis (θ) can be determined. The horizontal and vertical components are:

$$V_x = 15$$

$$V_y = 10$$

The magnitude V is given by:

$$V = \sqrt{V_x^2 + V_y^2}$$

and the angle θ is given by:

$$\theta = \tan^{-1}\left(\frac{V_y}{V_x}\right)$$

2.2.8   Vector Kinematics

Vector kinematics involves applying kinematic equations to vectors. For example, the displacement vector $\mathbf{d}$ can be related to initial velocity $\mathbf{v}_0$, acceleration $\mathbf{a}$, and time t using the equation:

$$\mathbf{d} = \mathbf{v}_0 t + \frac{1}{2}\mathbf{a}t^2$$

2.2.9   Example: Vector Kinematics

If an object starts at the origin with an initial velocity of $\mathbf{v}_0 = 5\hat{i} + 3\hat{j}$ and undergoes constant acceleration $\mathbf{a} = 2\hat{i} - \hat{j}$, the displacement vector $\mathbf{d}$ after $t = 4\,\mathrm{s}$ can be calculated using the vector kinematics equation.

2.2.10   Relative Motion in Two Dimensions

Relative motion in two dimensions involves analyzing the motion of one object with respect to another. The relative velocity $\mathbf{v}_{\mathrm{rel}}$ is the vector difference between the velocities of the two objects.

$$\mathbf{v}_{\mathrm{rel}} = \mathbf{v}_{\mathrm{B}} - \mathbf{v}_{\mathrm{A}}$$

2.2.11   Example: Relative Motion in Two Dimensions

If two cars A and B are moving with velocities $\mathbf{v}_{\mathrm{A}} = 20\hat{i} + 15\hat{j}$ and $\mathbf{v}_{\mathrm{B}} = 15\hat{i} - 10\hat{j}$, the relative velocity $\mathbf{v}_{\mathrm{rel}}$ of B with respect to A is calculated as:

$$\mathbf{v}_{\text{rel}} = (15 - 20)\hat{i} + (-10 - 15)\hat{j} = -5\hat{i} - 25\hat{j}$$

2.3 Projectile Motion

Projectile motion refers to the motion of an object projected into the air, influenced only by the force of gravity and air resistance (if present). This type of motion follows a curved path and can be described using the principles of kinematics.

2.3.1 Key Concepts

Launch Angle (θ)

The launch angle is the angle at which the projectile is initially launched with respect to the horizontal.

Initial Speed (v_0):

The initial speed is the magnitude of the velocity with which the projectile is launched.

Horizontal Motion

The horizontal motion of a projectile is uniform and can be described by the equation:

$$x = x_0 + v_{0x}t$$

where x_0 is the initial x-coordinate, v_{0x} is the initial x-component of velocity, and t is time.

Vertical Motion

The vertical motion of a projectile is uniformly accelerated due to gravity. The equations of motion are:

$$y = y_0 + v_{0y}t - \frac{1}{2}gt^2$$

$$v_y = v_{0y} - gt$$

where y_0 is the initial y-coordinate, v_{0y} is the initial y-component of velocity, g is the acceleration due to gravity, and v_y is the y-component of velocity.

2.3.2   Projectile Range

The range (R) of a projectile is the horizontal distance it travels before hitting the ground. For a projectile launched with an initial speed v_0 at an angle θ, the range is given by:

$$R = \frac{v_0^2 \sin(2\theta)}{g}$$

where g is the acceleration due to gravity.

2.3.3   Example: Projectile Range

Suppose a projectile is launched with an initial speed of $20\,\mathrm{m/s}$ at an angle of $45°$ above the horizontal. Using the range formula:

$$R = \frac{(20\,\mathrm{m/s})^2 \sin(2 \times 45°)}{9.8\,\mathrm{m/s}^2}$$

Solving for R:

$$R = \frac{400 \times 1}{9.8} \approx 40.82\,\mathrm{m}$$

2.3.4   Projectile Height

The maximum height (H) reached by a projectile is determined by its initial speed and launch angle. For a projectile launched with an initial speed v_0 at an angle θ, the maximum height is given by:

$$H = \frac{v_0^2 \sin^2(\theta)}{2g}$$

2.3.5 Example: Projectile Height

Using the same initial speed and launch angle as the previous example ($v_0 = 20\,\text{m/s}$, $\theta = 45°$), the maximum height is calculated using the height formula:

$$H = \frac{(20\,\text{m/s})^2 \sin^2(45°)}{2 \times 9.8\,\text{m/s}^2}$$

Solving for H:

$$H = \frac{400 \times 0.5}{19.6} \approx 10.20\,\text{m}$$

2.3.6 Projectile Time of Flight

The total time of flight (T) for a projectile is the duration it remains in the air. It is given by:

$$T = \frac{2v_{0y}}{g}$$

where v_{0y} is the initial y-component of velocity.

2.3.7 Example: Projectile Time of Flight

Continuing with the previous examples, the initial y-component of velocity (v_{0y}) can be calculated as:

$$v_{0y} = (20\,\text{m/s}) \sin(45°)$$

Substituting this into the time of flight formula:

$$T = \frac{2 \times (20\,\text{m/s}) \sin(45°)}{9.8\,\text{m/s}^2}$$

Solving for T:

$$T = \frac{40 \times 0.707}{9.8} \approx 2.88\,\text{s}$$

2.3.8   Projectile Motion with Air Resistance

In real-world scenarios, air resistance can impact projectile motion. However, the equations of motion become more complex in the presence of air resistance.

Chapter 3

Dynamics

3.1 Newton's Laws of Motion

Newton's laws of motion are fundamental principles that describe the relationship between the motion of an object and the forces acting on it. They form the basis for classical mechanics and are crucial for understanding the dynamics of objects.

3.1.1 First Law: Law of Inertia

Mathematical Formulation

Newton's first law states that an object at rest will remain at rest, and an object in motion will remain in motion with a constant velocity unless acted upon by a net external force. Mathematically, this can be expressed as:

$$\sum F = 0 \quad \text{(if velocity is constant)}$$

$$\sum F = m \cdot a \quad \text{(if acceleration is present)}$$

where $\sum F$ is the net force, m is the mass of the object, and a is the acceleration.

Example: Law of Inertia

Consider a car moving at a constant speed on a straight road. According to Newton's first law, the net force acting on the car is zero since there is no acceleration. If the car encounters an obstacle and comes to a stop, it is due to the application of external forces, such as friction and braking.

3.1.2 Second Law: Law of Acceleration

Mathematical Formulation

Newton's second law relates the net force acting on an object to its mass and acceleration. Mathematically, it is expressed as:

$$\sum F = m \cdot a$$

This law quantifies the concept of force and provides a way to calculate the resulting acceleration when a force is applied to an object.

Example: Law of Acceleration

If a force of $10\,\mathrm{N}$ is applied to an object with a mass of $2\,\mathrm{kg}$, the acceleration can be calculated using Newton's second law:

$$a = \frac{\sum F}{m} = \frac{10\,\mathrm{N}}{2\,\mathrm{kg}} = 5\,\mathrm{m/s}^2$$

3.1.3 Third Law: Action and Reaction

Mathematical Formulation

Newton's third law states that for every action, there is an equal and opposite reaction. Mathematically, if object A exerts a force $F_{\text{A on B}}$ on object B, then object B exerts a force $F_{\text{B on A}}$ on object A, and these forces are equal in magnitude but opposite in direction.

$$F_{\text{A on B}} = -F_{\text{B on A}}$$

Example: Action and Reaction

Consider a person standing on the ground. The person exerts a downward force on the ground (action), and by Newton's third law, the ground exerts an equal and opposite upward force on the person (reaction), preventing them from falling through the Earth.

3.1.4 Applications of Newton's Laws

Friction

Newton's laws are essential in understanding and calculating frictional forces. For example, the force of friction (F_{friction}) can be determined using the equation:

$$F_{\text{friction}} = \mu \cdot F_{\text{normal}}$$

where μ is the coefficient of friction and F_{normal} is the normal force.

Tension in Ropes

The tension (T) in a rope or cable can be analyzed using Newton's laws. For an object hanging vertically or moving on an inclined plane, the net force in the direction of the rope can be set equal to the tension.

$$T = m \cdot g \cdot \sin(\theta)$$

where θ is the angle of the incline.

Examples in Mechanics

Newton's laws find extensive applications in mechanics, including the study of planetary motion, satellite dynamics, and the behavior of fluids. These laws are foundational in classical mechanics and have stood the test of time.

3.1.5   Limitations of Newton's Laws

While Newton's laws are highly accurate for most everyday situations, they have limitations at very high speeds (near the speed of light) and extremely small scales (atomic and subatomic levels). In these cases, Einstein's theory of relativity and quantum mechanics are more appropriate.

3.2   Friction

Friction is a force that opposes the relative motion or tendency of such motion of two surfaces in contact. It plays a crucial role in everyday life, affecting the motion of objects on surfaces and influencing the efficiency of machines.

3.2.1   Types of Friction

Static Friction (F_{static}):

Static friction acts to prevent the initiation of motion between two surfaces. It is the force that must be overcome to set an object in motion.

$$F_{\text{static}} \leq \mu_{\text{static}} \cdot F_{\text{normal}}$$

where μ_{static} is the coefficient of static friction, and F_{normal} is the normal force.

Kinetic Friction (F_{kinetic}):

Kinetic friction acts when two surfaces are in relative motion. It is generally less than static friction and is given by:

$$F_{\text{kinetic}} = \mu_{\text{kinetic}} \cdot F_{\text{normal}}$$

where μ_{kinetic} is the coefficient of kinetic friction.

Rolling Friction

Rolling friction occurs when an object rolls over a surface. It is generally lower than sliding friction and depends on factors such as the shape and material of the rolling object.

3.2.2 Coefficient of Friction

The coefficient of friction (μ) is a dimensionless quantity that represents the ratio of the force of friction to the normal force between two surfaces.

$$\mu = \frac{F_{\text{friction}}}{F_{\text{normal}}}$$

3.2.3 Example: Calculating Frictional Force

Suppose a box with a mass of $10\,\text{kg}$ is placed on a horizontal surface with a coefficient of static friction (μ_{static}) of 0.5. The force needed to overcome static friction is:

$$F_{\text{static}} \leq \mu_{\text{static}} \cdot F_{\text{normal}}$$

Substituting values:

$$F_{\text{static}} \leq 0.5 \cdot (10\,\text{kg} \cdot 9.8\,\text{m/s}^2)$$

Solving for F_{static}:

$$F_{\text{static}} \leq 49\,\text{N}$$

3.2.4 Applications of Friction

Traction in Vehicles

Friction is crucial for providing traction in vehicles. The friction between tires and the road surface allows vehicles to accelerate, decelerate, and make turns safely.

Braking Systems

Friction is utilized in braking systems to slow down or stop moving objects. Brake pads create friction with the rotating wheels, converting kinetic energy into heat.

Belt and Pulley Systems

Friction is employed in belt and pulley systems to transmit motion and power between rotating shafts. The coefficient of friction is a crucial factor in the efficiency of such systems.

3.2.5 Challenges of Friction

While friction is essential in many applications, it also poses challenges. Excessive friction can lead to wear and tear on surfaces, reducing the efficiency of machines and causing energy losses.

3.2.6 Frictional Heating

When surfaces rub against each other, friction generates heat. This phenomenon is exploited in everyday activities, such as rubbing hands together to create warmth.

3.2.7 Coefficient of Friction in Inclined Planes

When a block is placed on an inclined plane, the coefficient of friction affects whether the block will slide or remain stationary. The force of friction is given by:

$$F_{\text{friction}} = \mu_{\text{kinetic}} \cdot F_{\text{normal}}$$

where F_{normal} is the normal force.

3.3   Circular Motion

Circular motion refers to the motion of an object along a circular path. This type of motion involves a continuous change in direction, and the object experiences centripetal acceleration directed toward the center of the circle.

3.3.1   Centripetal Acceleration

Centripetal acceleration (a_c) is the acceleration experienced by an object moving in a circular path. It is always directed toward the center of the circle and is given by:

$$a_c = \frac{v^2}{r}$$

where v is the tangential velocity of the object and r is the radius of the circular path.

3.3.2   Centripetal Force

Centripetal force (F_c) is the force required to keep an object moving in a circular path. It is related to centripetal acceleration by Newton's second law:

$$F_c = m \cdot a_c$$

where m is the mass of the object.

3.3.3   Example: Centripetal Acceleration

Suppose a car is moving around a curve with a radius of $20\,\mathrm{m}$ at a speed of $10\,\mathrm{m/s}$. The centripetal acceleration can be calculated using the formula:

$$a_c = \frac{v^2}{r} = \frac{(10\,\mathrm{m/s})^2}{20\,\mathrm{m}} = 5\,\mathrm{m/s}^2$$

3.3.4   Example: Centripetal Force

If the mass of the car is 1000 kg, the centripetal force can be calculated using Newton's second law:

$$F_c = m \cdot a_c = (1000\,\text{kg}) \cdot (5\,\text{m/s}^2) = 5000\,\text{N}$$

3.3.5   Angular Velocity and Frequency

Angular velocity (ω) is the rate at which an object rotates or revolves around a circle. It is related to linear velocity (v) and the radius (r) by the formula:

$$v = \omega \cdot r$$

Angular frequency (ω) is the number of revolutions per unit of time and is related to angular velocity by:

$$\omega = \frac{2\pi}{T}$$

where T is the period of one revolution.

3.3.6   Example: Angular Velocity

If a wheel completes one revolution in 2 s and has a radius of 0.5 m, the angular velocity (ω) can be calculated using the formula:

$$\omega = \frac{2\pi}{T} = \frac{2\pi}{2\,\text{s}} = \pi\,\text{rad/s}$$

3.3.7   Centrifugal Force

Centrifugal force is a fictitious force that appears to act on an object moving in a circular path, pushing it away from the center. It is equal in magnitude but opposite in direction to the centripetal force.

$$F_{\text{centrifugal}} = -F_c$$

3.3.8 Banked Curves

In banked curves, the road is tilted at an angle to the horizontal. This tilt provides the necessary centripetal force for a vehicle moving along the curve without relying solely on friction.

$$\tan(\theta) = \frac{v^2}{rg}$$

where θ is the angle of the bank, v is the speed of the vehicle, r is the radius of the curve, and g is the acceleration due to gravity.

3.3.9 Example: Banked Curve

Suppose a curve with a radius of 30 m is banked at an angle of 20°, and a car is moving along the curve at 15 m/s. The required banking angle can be calculated using the formula:

$$\tan(\theta) = \frac{v^2}{rg} = \frac{(15\,\text{m/s})^2}{30\,\text{m} \cdot 9.8\,\text{m/s}^2}$$

Solving for θ:

$$\theta = \tan^{-1}\left(\frac{(15\,\text{m/s})^2}{30\,\text{m} \cdot 9.8\,\text{m/s}^2}\right)$$

3.3.10 Applications of Circular Motion

Circular motion is prevalent in various real-world applications, including the motion of planets around the sun, the operation of amusement park rides, and the functioning of rotating machinery.

Chapter 4

Work, Energy, and Power

4.1 Work and Energy Theorems

Work and energy theorems are crucial concepts in physics that describe the transfer and transformation of energy in various physical systems.

4.1.1 Work (W)

Work is defined as the product of force (F) and displacement (d) in the direction of the force. Mathematically, it is given by:

$$W = F \cdot d \cdot \cos(\theta)$$

where θ is the angle between the force and the direction of displacement.

4.1.2 Example: Calculating Work

Suppose a force of $50\,\mathrm{N}$ is applied to move an object a distance of $10\,\mathrm{m}$ at an angle of $30°$ to the horizontal. The work done can be calculated using the formula:

$$W = F \cdot d \cdot \cos(\theta) = 50\,\mathrm{N} \cdot 10\,\mathrm{m} \cdot \cos(30°)$$

4.1.3   Kinetic Energy (KE)

Kinetic energy is the energy associated with the motion of an object. It is given by the formula:

$$KE = \frac{1}{2}mv^2$$

where m is the mass of the object and v is its velocity.

4.1.4   Example: Calculating Kinetic Energy

If an object with a mass of 2 kg is moving with a velocity of 4 m/s, the kinetic energy can be calculated using the formula:

$$KE = \frac{1}{2}mv^2 = \frac{1}{2} \cdot 2\,\text{kg} \cdot (4\,\text{m/s})^2$$

4.1.5   Potential Energy (PE)

Potential energy is the energy stored in an object due to its position or state. For gravitational potential energy, it is given by:

$$PE = mgh$$

where m is the mass, g is the acceleration due to gravity, and h is the height above a reference point.

4.1.6   Example: Calculating Gravitational Potential Energy

If an object with a mass of 5 kg is lifted to a height of 10 m above the ground, the gravitational potential energy can be calculated using the formula:

$$PE = mgh = 5\,\text{kg} \cdot 9.8\,\text{m/s}^2 \cdot 10\,\text{m}$$

4.1.7 Work-Energy Theorem

The work-energy theorem states that the work done on an object is equal to the change in its kinetic energy. Mathematically, it is expressed as:

$$W = \Delta KE$$

This theorem is a powerful tool for analyzing the motion of objects subjected to various forces.

4.1.8 Example: Work-Energy Theorem

Suppose a force of 30 N is applied to an object, causing it to accelerate from rest to a velocity of 6 m/s over a distance of 5 m. The work done can be calculated, and using the work-energy theorem:

$$W = \Delta KE$$

$$30\,\text{N} \cdot 5\,\text{m} = \frac{1}{2}m(6\,\text{m/s})^2$$

Solving for m, the mass of the object.

4.1.9 Conservative and Non-conservative Forces

In the context of energy, forces are classified as conservative or non-conservative. Conservative forces, such as gravity, do work that is path-independent, and the total mechanical energy (kinetic + potential) of an object is conserved. Non-conservative forces, like friction, result in energy dissipation.

4.1.10 Example: Conservative Force

A ball rolling down a frictionless hill experiences only gravitational force, which is a conservative force. The total mechanical energy of the ball is conserved as it moves.

4.1.11   Power (P)

Power is the rate at which work is done or energy is transferred. It is given by the formula:

$$P = \frac{W}{\Delta t}$$

where W is the work done, and Δt is the time taken.

4.1.12   Example: Calculating Power

If $1000\,\mathrm{J}$ of work is done in $10\,\mathrm{s}$, the power can be calculated using the formula:

$$P = \frac{W}{\Delta t} = \frac{1000\,\mathrm{J}}{10\,\mathrm{s}}$$

4.1.13   Applications of Work and Energy

The concepts of work and energy are applied in various fields, from understanding the motion of celestial bodies to designing efficient machines and analyzing collisions.

4.2   Conservation of Mechanical Energy

The conservation of mechanical energy is a fundamental principle in physics, stating that the total mechanical energy of a system remains constant if only conservative forces, such as gravity, are acting on it. This principle is derived from the work-energy theorem and is expressed mathematically as:

$$E_{\mathrm{mechanical}} = KE + PE = \mathrm{constant}$$

where $E_{\mathrm{mechanical}}$ is the total mechanical energy, KE is the kinetic energy, and PE is the potential energy.

4.2.1   Example: Conservation of Mechanical Energy

Consider a pendulum of mass m released from a certain height h. As the pendulum swings, its potential energy is converted into kinetic energy and vice versa. At any point during the motion, the total mechanical energy remains constant.

$$E_{\text{mechanical}} = KE + PE$$

$$mgh = \frac{1}{2}mv^2 + mgh'$$

where v is the velocity at the lowest point, and h' is the height at that point.

4.2.2   Application: Roller Coaster

A classic example of the conservation of mechanical energy is a roller coaster. As the coaster moves along the track, gravitational potential energy is converted into kinetic energy and vice versa. At any point, the sum of kinetic and potential energy remains constant, neglecting energy losses due to non-conservative forces like air resistance.

$$E_{\text{mechanical}} = KE + PE$$

$$mgh = \frac{1}{2}mv^2 + mgh'$$

4.2.3   Spring Potential Energy

In addition to gravitational potential energy, systems involving springs also exhibit potential energy. The potential energy stored in a spring (PE_{spring}) is given by:

$$PE_{\text{spring}} = \frac{1}{2}kx^2$$

where k is the spring constant, and x is the displacement from the equilibrium position.

4.2.4   Example: Spring System

Consider a mass m attached to a spring with spring constant k. If the spring is compressed by a distance x from its equilibrium position, the potential energy stored in the spring can be calculated using the formula:

$$PE_{\text{spring}} = \frac{1}{2}kx^2$$

4.2.5   Conservation of Mechanical Energy in Horizontal Motion

When only conservative forces are at play, the conservation of mechanical energy holds even in horizontal motion. The kinetic energy and potential energy due to height changes are the primary contributors to the total mechanical energy.

$$E_{\text{mechanical}} = KE + PE$$

$$\frac{1}{2}mv^2 + PE_{\text{height}} = \text{constant}$$

4.2.6   Example: Horizontal Motion

Consider a block sliding on a frictionless surface. If the block starts from rest at a certain height and slides down an incline, the conservation of mechanical energy can be applied.

$$\frac{1}{2}mv^2 + mgh = \text{constant}$$

4.2.7 Limitations of Conservation of Mechanical Energy

While the conservation of mechanical energy is a powerful tool in many situations, it is important to note its limitations. It assumes the absence of non-conservative forces, such as air resistance or friction. In real-world scenarios, energy losses to these forces may occur.

4.2.8 Example: Losses in Mechanical Energy

In a swinging pendulum, air resistance and internal friction within the support may lead to energy losses. Over time, the amplitude of the pendulum's swing decreases, and its total mechanical energy is not conserved.

4.3 Power and its Application

Power is a fundamental concept in physics, representing the rate at which work is done or energy is transferred. It is a crucial parameter in various applications, from understanding the performance of engines to designing efficient electrical systems.

4.3.1 Power (P)

Power is defined as the rate of doing work or the rate of energy transfer. Mathematically, it is expressed as:

$$P = \frac{W}{\Delta t}$$

where P is power, W is the work done, and Δt is the time taken.

4.3.2 Example: Calculating Power

Suppose a force of $100\,\text{N}$ is applied to lift a box vertically by $2\,\text{m}$ in $5\,\text{s}$. The work done can be calculated using the formula:

$$W = F \cdot d \cdot \cos(\theta)$$

And then, the power can be calculated using the formula $P = \frac{W}{\Delta t}$.

$$P = \frac{F \cdot d \cdot \cos(\theta)}{\Delta t}$$

4.3.3   Unit of Power

The standard unit of power in the International System of Units (SI) is the watt (W), where $1\,\mathrm{W} = 1\,\mathrm{J/s}$. Other common units include kilowatts (kW) and horsepower (hp).

4.3.4   Applications of Power

Mechanical Power in Machines

In mechanical systems, power is essential for describing the performance of engines and machines. The power output of an engine, for example, determines its ability to perform work.

$$P = \frac{W}{\Delta t}$$

Electrical Power in Circuits

In electrical systems, power is the rate of energy transfer in circuits. The power (P) in an electrical circuit can be calculated using Ohm's Law:

$$P = I \cdot V$$

where I is the current and V is the voltage.

4.3.5 Example: Electrical Power

Suppose a device operates with a current of $5\,\text{A}$ and a voltage of $10\,\text{V}$. The electrical power consumed by the device can be calculated using the formula $P = I \cdot V$.

$$P = 5\,\text{A} \cdot 10\,\text{V}$$

4.3.6 Thermal Power in Heat Engines

In heat engines, power is associated with the rate at which heat energy is converted into mechanical work. The efficiency of a heat engine is defined as the ratio of the useful work output to the heat input.

$$\text{Efficiency} = \frac{\text{Useful Work Output}}{\text{Heat Input}} \times 100\%$$

4.3.7 Example: Heat Engine Efficiency

Consider a heat engine that produces $2000\,\text{J}$ of useful work output while absorbing $4000\,\text{J}$ of heat from a high-temperature reservoir. The efficiency can be calculated using the formula:

$$\text{Efficiency} = \frac{\text{Useful Work Output}}{\text{Heat Input}} \times 100\%$$

$$\text{Efficiency} = \frac{2000\,\text{J}}{4000\,\text{J}} \times 100\%$$

4.3.8 Power in Physics and Sports

In sports, power is a crucial factor, especially in activities requiring explosive movements. The power generated by an athlete can be calculated using the formula:

$$P = \frac{\Delta \text{Work}}{\Delta t}$$

where ΔWork is the change in work done, and Δt is the corresponding time interval.

4.3.9   Example: Power in Sports

Suppose a weightlifter lifts a 100 kg barbell from the ground to a height of 2 m in 2 s. The power generated by the weightlifter can be calculated using the formula:

$$P = \frac{F \cdot d \cdot \cos(\theta)}{\Delta t}$$

Chapter 5

Momentum and Collisions

5.1 Impulse and Linear Momentum

Impulse and linear momentum are crucial concepts in physics, providing insights into the motion of objects and the effects of forces on their trajectories.

5.1.1 Linear Momentum (p)

Linear momentum is the product of an object's mass (m) and its velocity (v). Mathematically, it is expressed as:

$$p = m \cdot v$$

The SI unit of linear momentum is $\text{kg} \cdot \text{m/s}$.

5.1.2 Example: Calculating Linear Momentum

Consider a car with a mass of $1500\,\text{kg}$ traveling at a velocity of $20\,\text{m/s}$. The linear momentum of the car can be calculated using the formula $p = m \cdot v$.

$$p = 1500\,\text{kg} \cdot 20\,\text{m/s}$$

51

5.1.3   Impulse (J)

Impulse is the product of force (F) and the time (Δt) over which the force is applied. Mathematically, it is expressed as:

$$J = F \cdot \Delta t$$

Impulse is also equal to the change in linear momentum of an object.

$$J = \Delta p$$

5.1.4   Example: Calculating Impulse

If a force of $300\,\mathrm{N}$ is applied to an object for $5\,\mathrm{s}$, the impulse can be calculated using the formula $J = F \cdot \Delta t$.

$$J = 300\,\mathrm{N} \cdot 5\,\mathrm{s}$$

5.1.5   Conservation of Linear Momentum

In a closed system, the total linear momentum remains constant if no external forces act on the system. This principle is known as the conservation of linear momentum.

$$\text{Total Initial Momentum} = \text{Total Final Momentum}$$

5.1.6   Example: Conservation of Linear Momentum

Consider two ice skaters initially at rest. If they push off each other, the total linear momentum before the push is zero, and after the push, it remains zero. This demonstrates the conservation of linear momentum.

5.1.7 Collisions and Types of Collisions

Collisions involve interactions between objects, and they are categorized as elastic or inelastic.

Elastic Collisions

In elastic collisions, both kinetic energy and linear momentum are conserved.

$$\text{Total Initial Momentum} = \text{Total Final Momentum}$$

$$\text{Total Initial Kinetic Energy} = \text{Total Final Kinetic Energy}$$

Inelastic Collisions

In inelastic collisions, only linear momentum is conserved, while kinetic energy is not.

$$\text{Total Initial Momentum} = \text{Total Final Momentum}$$

$$\text{Total Initial Kinetic Energy} \neq \text{Total Final Kinetic Energy}$$

5.1.8 Example: Elastic Collision

Two billiard balls collide elastically on a frictionless table. The total linear momentum and kinetic energy before and after the collision are conserved.

$$\text{Total Initial Momentum} = \text{Total Final Momentum}$$

$$\text{Total Initial Kinetic Energy} = \text{Total Final Kinetic Energy}$$

5.1.9   Example: Inelastic Collision

Two objects stick together after colliding on a frictionless surface. In this inelastic collision, the total linear momentum is conserved, but the total kinetic energy is not.

$$\text{Total Initial Momentum} = \text{Total Final Momentum}$$

$$\text{Total Initial Kinetic Energy} \neq \text{Total Final Kinetic Energy}$$

5.1.10   Applications of Impulse and Linear Momentum

Understanding impulse and linear momentum is crucial in analyzing the motion of objects, designing safety features in vehicles, and predicting the outcomes of collisions in various scenarios.

5.2   Collisions and Conservation of Momentum

Collisions are fundamental events in physics, involving interactions between objects. The conservation of linear momentum is a crucial principle in understanding and analyzing the outcomes of these collisions.

5.2.1   Types of Collisions

Collisions can be categorized into two main types: elastic and inelastic.

Elastic Collisions

In elastic collisions, both linear momentum and kinetic energy are conserved. The total initial momentum of the system is equal to the total final momentum.

$$\text{Total Initial Momentum} = \text{Total Final Momentum}$$

Total Initial Kinetic Energy = Total Final Kinetic Energy

Inelastic Collisions

In inelastic collisions, only linear momentum is conserved. While the total initial momentum is equal to the total final momentum, the total initial kinetic energy is not necessarily equal to the total final kinetic energy.

Total Initial Momentum = Total Final Momentum

Total Initial Kinetic Energy $\neq$ Total Final Kinetic Energy

5.2.2 Example: Elastic Collision

Consider two billiard balls of masses m_1 and m_2 colliding elastically. The conservation of linear momentum and kinetic energy can be expressed as:

$$m_1 u_1 + m_2 u_2 = m_1 v_1 + m_2 v_2$$

$$\frac{1}{2} m_1 u_1^2 + \frac{1}{2} m_2 u_2^2 = \frac{1}{2} m_1 v_1^2 + \frac{1}{2} m_2 v_2^2$$

where u_1 and u_2 are the initial velocities, and v_1 and v_2 are the final velocities of the two balls.

5.2.3 Example: Inelastic Collision

Consider two objects colliding and sticking together. The conservation of linear momentum in an inelastic collision is expressed as:

$$m_1 u_1 + m_2 u_2 = (m_1 + m_2) V_f$$

where V_f is the final velocity of the combined mass.

5.2.4 Applications of Conservation of Momentum

The conservation of momentum plays a crucial role in various real-world applications.

Vehicle Collisions

In car collisions, the principles of momentum conservation are used to analyze the forces involved and design safety features to reduce the impact on passengers.

Sports

In sports like football or soccer, understanding the conservation of momentum helps in predicting the trajectories of moving objects and optimizing players' strategies.

Rocket Propulsion

In rocket propulsion, the ejection of exhaust gases leads to a change in the rocket's momentum, demonstrating the conservation of linear momentum.

5.2.5 Example: Rocket Propulsion

Consider a rocket of mass M ejecting exhaust gases with velocity v_e. The change in momentum (Δp) of the rocket can be calculated using:

$$\Delta p = M \cdot v_e$$

This change in momentum is equal and opposite to the change in momentum of the ejected gases.

Chapter 6

Rotational Motion

6.1 Rotational Kinematics and Dynamics

Rotational motion involves the motion of objects around a fixed axis. Understanding rotational kinematics and dynamics is essential for describing the behavior of rotating objects.

6.1.1 Rotational Kinematics

Angular Displacement (θ)

Angular displacement is the measure of the angle through which an object rotates. It is denoted by θ and is measured in radians.

$$\theta = \frac{s}{r}$$

where s is the linear displacement and r is the radius.

Angular Velocity (ω)

Angular velocity is the rate at which an object rotates. It is given by the ratio of angular displacement (θ) to the change in time (Δt).

$$\omega = \frac{\Delta\theta}{\Delta t}$$

Angular Acceleration (α)

Angular acceleration is the rate at which angular velocity changes with time. It is given by the ratio of the change in angular velocity ($\Delta\omega$) to the change in time (Δt).

$$\alpha = \frac{\Delta\omega}{\Delta t}$$

6.1.2   Example: Rotational Kinematics

Consider a wheel with a radius of $0.5\,\text{m}$ rotating for $4\,\text{s}$ with an initial angular velocity of $2\,\text{rad/s}$. The angular displacement (θ) can be calculated using the formula $\theta = \omega_0 \cdot t + \frac{1}{2}\alpha t^2$.

$$\theta = (2\,\text{rad/s} \cdot 4\,\text{s}) + \frac{1}{2}(0) \cdot (4\,\text{s})^2$$

6.1.3   Rotational Dynamics

Torque (τ)

Torque is the rotational equivalent of force and causes angular acceleration. It is given by the product of the force (F) applied perpendicular to the lever arm (r).

$$\tau = r \cdot F$$

Rotational Inertia (I)

Rotational inertia is the measure of an object's resistance to changes in its rotational motion. It depends on both the mass distribution and the axis of rotation.

$$I = \sum m_i \cdot r_i^2$$

where m_i is the mass of each particle, and r_i is its perpendicular distance from the axis of rotation.

Angular Momentum (L)

Angular momentum is the rotational equivalent of linear momentum and is given by the product of rotational inertia (I) and angular velocity (ω).

$$L = I \cdot \omega$$

6.1.4 Example: Rotational Dynamics

Consider a door with a rotational inertia (I) of $3\,\mathrm{kg \cdot m^2}$ and a person applying a force (F) of $10\,\mathrm{N}$ at a distance (r) of $0.8\,\mathrm{m}$ from the hinge. The torque (τ) applied can be calculated using the formula $\tau = r \cdot F$.

$$\tau = 0.8\,\mathrm{m} \cdot 10\,\mathrm{N}$$

6.1.5 Conservation of Angular Momentum

In the absence of external torques, the conservation of angular momentum states that the total angular momentum of a system remains constant.

Total Initial Angular Momentum = Total Final Angular Momentum

6.1.6 Example: Conservation of Angular Momentum

Consider an ice skater spinning with arms extended. When the skater brings arms closer to the body, the moment of inertia decreases, causing an increase in angular velocity, thus conserving angular momentum.

6.2   Torque and Angular Momentum

Torque and angular momentum are crucial concepts in understanding the rotational motion of objects. They provide insights into the causes and effects of rotational motion.

6.2.1   Torque (τ)

Torque is the rotational equivalent of force. It is the measure of the tendency of a force to rotate an object about an axis. Mathematically, it is given by:

$$\tau = r \cdot F$$

where τ is the torque, r is the lever arm (distance from the axis of rotation), and F is the force applied perpendicular to the lever arm.

6.2.2   Example: Calculating Torque

Consider a wrench applying a force of $20\,\mathrm{N}$ at a distance of $0.3\,\mathrm{m}$ from the axis of rotation. The torque applied can be calculated using the formula $\tau = r \cdot F$.

$$\tau = 0.3\,\mathrm{m} \cdot 20\,\mathrm{N}$$

6.2.3   Rotational Inertia (I)

Rotational inertia is the measure of an object's resistance to changes in its rotational motion. It depends on both the mass distribution and the axis of rotation. The formula for rotational inertia is given by:

$$I = \sum m_i \cdot r_i^2$$

where m_i is the mass of each particle, and r_i is its perpendicular distance from the axis of rotation.

6.2.4 Example: Calculating Rotational Inertia

Consider a disc of radius 0.4 m and mass 2 kg rotating about its central axis. The rotational inertia of the disc can be calculated using the formula $I = \frac{1}{2}mr^2$.

$$I = \frac{1}{2} \cdot 2\,\text{kg} \cdot (0.4\,\text{m})^2$$

6.2.5 Angular Momentum (L)

Angular momentum is the rotational equivalent of linear momentum. It is given by the product of rotational inertia (I) and angular velocity (ω).

$$L = I \cdot \omega$$

6.2.6 Example: Calculating Angular Momentum

Consider a wheel with a rotational inertia (I) of 1 kg·m^2 and an angular velocity (ω) of 5 rad/s. The angular momentum (L) can be calculated using the formula $L = I \cdot \omega$.

$$L = 1\,\text{kg} \cdot \text{m}^2 \cdot 5\,\text{rad/s}$$

6.2.7 Conservation of Angular Momentum

In the absence of external torques, the conservation of angular momentum states that the total angular momentum of a system remains constant.

$$\text{Total Initial Angular Momentum} = \text{Total Final Angular Momentum}$$

6.2.8 Example: Conservation of Angular Momentum

Consider an ice skater spinning with arms extended. When the skater brings arms closer to the body, the moment of inertia decreases, causing an increase in angular velocity, thus conserving angular momentum.

6.2.9   Applications of Torque and Angular Momentum

Understanding torque and angular momentum is crucial in various real-world applications, from designing rotational machinery to explaining celestial phenomena.

Rotational Machinery

In the design of gears, flywheels, and other rotational machinery, torque and angular momentum calculations are essential for optimal performance.

Celestial Phenomena

In celestial mechanics, the conservation of angular momentum helps explain the rotational motion of planets and other celestial bodies.

Sports

In sports like figure skating or gymnastics, athletes manipulate their body shape to control angular momentum, enabling them to perform complex maneuvers.

Chapter 7

Thermodynamics

7.1 Temperature and Heat

Temperature and heat are fundamental concepts in thermodynamics, providing a basis for understanding the behavior of substances in different conditions.

7.1.1 Temperature (T)

Temperature is a measure of the average kinetic energy of the particles in a substance. The Celsius ($°C$) and Kelvin (K) scales are commonly used to express temperature.

Conversion between Celsius and Kelvin

The conversion between Celsius and Kelvin is given by:

$$T(\text{K}) = T(°\text{C}) + 273.15$$

7.1.2 Example: Temperature Conversion

If the temperature is given as $20\,°C$, the equivalent temperature in Kelvin (T) can be calculated as:

$$T = 20 + 273.15$$

7.1.3 Heat (Q)

Heat is the transfer of thermal energy between systems. It is measured in joules (J).

Heat Transfer Equation

The amount of heat transferred (Q) is related to the mass (m) of the substance, its specific heat capacity (c), and the change in temperature (ΔT) by the equation:

$$Q = mc\Delta T$$

7.1.4 Example: Heat Transfer

If $500\,\text{g}$ of water ($c = 4.18\,\text{J/g}^\circ\text{C}$) is heated from $20\,^\circ\text{C}$ to $50\,^\circ\text{C}$, the amount of heat transferred (Q) can be calculated using the formula $Q = mc\Delta T$.

$$Q = (500\,\text{g}) \cdot (4.18\,\text{J/g}^\circ\text{C}) \cdot (50 - 20)$$

7.1.5 Specific Heat Capacity (c)

Specific heat capacity is the amount of heat required to raise the temperature of one unit mass of a substance by one degree Celsius. It is measured in $\text{J/g}^\circ\text{C}$.

Water's Specific Heat Capacity

For water, the specific heat capacity (c) is approximately $4.18\,\text{J/g}^\circ\text{C}$.

7.1.6 Example: Specific Heat Capacity

If $100\,\text{g}$ of aluminum ($c = 0.9\,\text{J/g}^\circ\text{C}$) is heated, the amount of heat (Q) required to raise its temperature by $30\,^\circ\text{C}$ can be calculated using the formula $Q = mc\Delta T$.

$$Q = (100\,\text{g}) \cdot (0.9\,\text{J/g}^\circ\text{C}) \cdot (30)$$

7.1.7 Latent Heat (L)

Latent heat is the heat absorbed or released during a phase change (e.g., melting or boiling). It is given by:

$$Q = mL$$

where m is the mass of the substance and L is the latent heat.

7.1.8 Example: Latent Heat of Fusion

If $200\,\text{g}$ of ice is melted at $0\,^\circ\text{C}$ ($L = 334\,\text{J/g}$), the amount of heat absorbed (Q) can be calculated using the formula $Q = mL$.

$$Q = (200\,\text{g}) \cdot (334\,\text{J/g})$$

7.1.9 Applications of Temperature and Heat

Understanding temperature and heat is crucial in various applications, from designing thermal systems to predicting climate changes.

Thermal Comfort in Buildings

In building design, knowledge of heat transfer is essential for creating comfortable living or working environments.

Climate Science

In climate science, the study of heat transfer helps scientists model and predict global climate patterns.

Thermal Engineering

In thermal engineering, understanding temperature and heat is fundamental to designing efficient heating, ventilation, and air conditioning (HVAC) systems.

7.2   Laws of Thermodynamics

The laws of thermodynamics govern the behavior of energy and matter in physical systems. They provide a framework for understanding the principles that govern heat, work, and the direction of physical processes.

7.2.1   Zeroth Law of Thermodynamics

The Zeroth Law states that if two systems are each in thermal equilibrium with a third system, then they are in thermal equilibrium with each other.

Temperature Measurement

The concept of temperature is based on the Zeroth Law, allowing the definition of a temperature scale.

Example: Zeroth Law

If two objects, A and B, are in thermal equilibrium with a third object, C, then objects A and B are in thermal equilibrium with each other.

7.2.2   First Law of Thermodynamics

The First Law, also known as the Law of Energy Conservation, states that energy cannot be created or destroyed in an isolated system; it can only change forms.

First Law Equation

For a thermodynamic process, the equation is given by:

$$\Delta U = Q - W$$

where ΔU is the change in internal energy, Q is the heat added to the system, and W is the work done by the system.

Example: First Law

Consider a gas undergoing an isothermal expansion ($\Delta T = 0$). The work done (W) can be calculated using the equation $W = P \cdot \Delta V$.

$$\Delta U = Q - P \cdot \Delta V$$

7.2.3 Second Law of Thermodynamics

The Second Law introduces the concept of entropy, stating that in any energy transfer or transformation, if no energy enters or leaves the system, the potential energy of the state will always be less than that of the initial state.

Entropy Increase

The entropy of an isolated system tends to increase over time.

Example: Second Law

Consider the irreversible process of mixing hot and cold water. The entropy of the system increases as the two temperatures approach equilibrium.

7.2.4 Third Law of Thermodynamics

The Third Law states that the entropy of a perfect crystal approaches zero as the temperature approaches absolute zero ($T \to 0$).

Absolute Zero Entropy

At absolute zero, a perfect crystal has minimal entropy.

Example: Third Law

As the temperature of a system decreases toward absolute zero, the entropy approaches a minimum value.

7.2.5   Applications of Thermodynamic Laws

Understanding the laws of thermodynamics is essential in various applications, ranging from heat engines to refrigerators and the behavior of gases.

Heat Engines

In designing heat engines, the efficiency is constrained by the Second Law of Thermodynamics.

Refrigerators

The operation of refrigerators is governed by the principles of thermodynamics, particularly the transfer of heat.

Gaseous Systems

In studying the behavior of gases, thermodynamic laws help explain phenomena such as temperature changes during expansion or compression.

7.3   Kinetic Theory of Gases

The kinetic theory of gases provides a microscopic understanding of the behavior of gases, explaining macroscopic properties such as pressure, temperature, and volume in terms of the motion of individual gas particles.

7.3.1 Assumptions of Kinetic Theory

The kinetic theory makes several assumptions about ideal gases:

1. Gas particles are in constant random motion. 2. The volume occupied by gas particles is negligible compared to the total volume. 3. Gas particles undergo elastic collisions with each other and with the container walls. 4. There are no attractive or repulsive forces between gas particles.

7.3.2 Ideal Gas Law

The ideal gas law relates the pressure (P), volume (V), and temperature (T) of an ideal gas. It is given by:

$$PV = nRT$$

where n is the number of moles of gas and R is the ideal gas constant.

Example: Ideal Gas Law

Consider a container with 2 moles of gas at a temperature of 300 K and a volume of 10 L. The pressure (P) can be calculated using the ideal gas law.

$$P = \frac{nRT}{V}$$

7.3.3 Mean Free Path (λ)

The mean free path is the average distance a gas particle travels between collisions. It is given by:

$$\lambda = \frac{kT}{\sqrt{2}\pi d^2 P}$$

where k is the Boltzmann constant, T is the temperature, d is the diameter of a gas molecule, and P is the pressure.

Example: Mean Free Path

For a gas at $300\,\mathrm{K}$, $P = 1\,\mathrm{atm}$, and $d = 3\,\text{Å}$, the mean free path can be calculated.

$$\lambda = \frac{k \cdot 300}{\sqrt{2}\pi \cdot (3 \times 10^{-10})^2 \cdot 1\,\mathrm{atm}}$$

Root Mean Square Speed (v_{rms})

The root mean square speed is the square root of the average of the squares of the individual gas particle speeds. It is given by:

$$v_{\mathrm{rms}} = \sqrt{\frac{3kT}{m}}$$

where m is the mass of a gas molecule.

Example: Root Mean Square Speed

For a gas at $300\,\mathrm{K}$ with a molecular mass of $2\,\mathrm{g/mol}$, the root mean square speed can be calculated.

$$v_{\mathrm{rms}} = \sqrt{\frac{3k \cdot 300}{2\,\mathrm{g/mol}}}$$

7.3.4   Maxwell-Boltzmann Distribution

The Maxwell-Boltzmann distribution describes the distribution of speeds for gas particles at a particular temperature. It is given by the probability density function:

$$f(v) = 4\pi \left(\frac{m}{2\pi kT}\right)^{3/2} v^2 e^{-\frac{mv^2}{2kT}}$$

Example: Maxwell-Boltzmann Distribution

The probability of finding a gas particle with a speed between v_1 and v_2 can be calculated using the Maxwell-Boltzmann distribution.

$$\int_{v_1}^{v_2} f(v) \, dv$$

7.3.5 Applications of Kinetic Theory

Understanding the kinetic theory of gases is essential in various applications, including the design of gas-based technologies and the study of planetary atmospheres.

Gas Technologies

In the design of gas-based technologies such as engines and turbines, knowledge of gas behavior is crucial.

Planetary Atmospheres

Studying the kinetic theory helps in understanding the composition and behavior of planetary atmospheres.

Thermal Conductivity

The kinetic theory also contributes to explaining thermal conductivity in gases.

Chapter 8

Waves and Optics

8.1 Wave Properties

Waves are fundamental phenomena in physics, exhibiting various properties that govern their behavior. Understanding these properties is crucial in fields ranging from optics to acoustics and beyond.

8.1.1 Wave Equation

The general form of a wave equation is given by:

$$y(x,t) = A\sin(kx - \omega t)$$

where y is the displacement of the wave, A is the amplitude, k is the wave number, x is the position, ω is the angular frequency, and t is the time.

Example: Simple Harmonic Wave

Consider a simple harmonic wave with an amplitude of $2\,\text{cm}$, wave number $k = 0.1\,\text{rad/m}$, and angular frequency $\omega = 2\,\text{rad/s}$. The wave equation becomes:

$$y(x,t) = 2\sin(0.1x - 2t)$$

73

8.1.2   Wave Speed

The wave speed (v) is the rate at which a wave propagates through a medium.
It is given by the equation:

$$v = \frac{\lambda}{T}$$

where λ is the wavelength and T is the period of the wave.

Example: Wave Speed

For a wave with a wavelength of 5 m and a period of 2 s, the wave speed can be
calculated using the formula $v = \frac{\lambda}{T}$.

$$v = \frac{5\,\text{m}}{2\,\text{s}}$$

8.1.3   Wavelength (λ)

The wavelength is the distance between two successive points in a wave that are
in phase. It is related to the wave speed and frequency by the equation:

$$\lambda = \frac{v}{f}$$

where v is the wave speed and f is the frequency.

Example: Wavelength

If a wave with a frequency of 50 Hz travels with a speed of 300 m/s, the wavelength can be calculated using $\lambda = \frac{v}{f}$.

$$\lambda = \frac{300\,\text{m/s}}{50\,\text{Hz}}$$

8.1.4 Wave Interference

Wave interference occurs when two or more waves overlap. It can result in constructive interference (amplitude increases) or destructive interference (amplitude decreases).

Example: Interference Patterns

Consider two waves with equal amplitudes (A) and wavelengths (λ) traveling in the same direction. Constructive interference occurs when $A_1 + A_2$, and destructive interference occurs when $A_1 - A_2$.

$$y(x,t) = A\sin(kx - \omega t) + A\sin(kx - \omega t + \phi)$$

8.1.5 Diffraction

Diffraction is the bending of waves around obstacles or the spreading of waves when passing through narrow openings. It is characterized by the diffraction angle (θ).

Example: Diffraction Pattern

For a wave passing through a single slit, the diffraction pattern can be described by the equation $a\sin(\theta) = m\lambda$, where a is the slit width, θ is the diffraction angle, m is the order of the diffraction maximum, and λ is the wavelength.

8.1.6 Applications of Wave Properties

Understanding wave properties is crucial in various applications, from designing communication systems to explaining optical phenomena.

Communication Systems

In designing communication systems, knowledge of wave properties is essential for signal transmission and reception.

Optical Phenomena

Explaining optical phenomena, such as interference and diffraction, requires a thorough understanding of wave properties.

8.2 Geometric Optics

Geometric optics is a branch of optics that describes the behavior of light in terms of rays. This approach is particularly useful when dealing with lenses, mirrors, and other optical elements.

8.2.1 Reflection

The law of reflection states that the angle of incidence (θ_{in}) is equal to the angle of reflection (θ_{re}), and both angles are measured with respect to the normal.

$$\theta_{\text{in}} = \theta_{\text{re}}$$

Example: Reflection

Consider a light ray incident on a mirror at an angle of 30°. According to the law of reflection, the reflected ray will also make an angle of 30° with the normal.

$$\theta_{\text{in}} = \theta_{\text{re}} = 30°$$

8.2.2 Refraction

The law of refraction (Snell's Law) relates the angles of incidence and refraction to the indices of refraction of the two media involved. It is given by:

$$n_1 \sin(\theta_{\text{in}}) = n_2 \sin(\theta_{\text{re}})$$

where n_1 and n_2 are the refractive indices of the first and second media, respectively.

Example: Refraction

If light passes from air $(n_1 = 1)$ into water $(n_2 = 1.33)$ and strikes the surface at $45°$, the angle of refraction can be calculated using Snell's Law.

$$n_1 \sin(\theta_{\text{in}}) = n_2 \sin(\theta_{\text{re}})$$

8.2.3 Lens Equation

The lens equation relates the object distance (d_o), the image distance (d_i), and the focal length (f) of a lens:

$$\frac{1}{f} = \frac{1}{d_o} + \frac{1}{d_i}$$

Example: Lens Equation

For a converging lens with a focal length of $10\,\text{cm}$ and an object placed $20\,\text{cm}$ away, the image distance can be calculated using the lens equation.

$$\frac{1}{f} = \frac{1}{d_o} + \frac{1}{d_i}$$

8.2.4 Magnification

The magnification (m) of an optical system is given by the ratio of the image height (h_i) to the object height (h_o):

$$m = \frac{h_i}{h_o} = -\frac{d_i}{d_o}$$

Example: Magnification

If an object is placed $15\,\text{cm}$ from a diverging lens with a focal length of $-10\,\text{cm}$, the magnification can be calculated.

$$m = -\frac{d_i}{d_o}$$

8.2.5 Applications of Geometric Optics

Geometric optics is widely used in designing optical instruments such as cameras, microscopes, and telescopes.

Camera Design

Understanding reflection and refraction is essential in designing camera lenses for capturing images.

Microscope Operation

Geometric optics principles are applied in the design of microscopes, allowing for magnification and clear visualization of microscopic objects.

Telescope Functionality

Telescopes utilize lenses and mirrors based on geometric optics principles to observe distant celestial objects.

8.3 Wave Optics

Wave optics, also known as physical optics, focuses on the wave nature of light. It explores phenomena such as interference, diffraction, and polarization, providing a comprehensive understanding of light behavior.

8.3.1 Young's Double-Slit Experiment

Young's double-slit experiment demonstrates the interference of light waves. The fringe separation (d) is given by:

$$d = \frac{\lambda L}{a}$$

where λ is the wavelength of light, L is the distance from the slits to the screen, and a is the slit separation.

Example: Double-Slit Interference

Consider a double-slit experiment with a slit separation of 0.1 mm, a wavelength of 500 nm, and a screen distance of 2 m. The fringe separation can be calculated using the formula $d = \frac{\lambda L}{a}$.

$$d = \frac{(500 \times 10^{-9}) \times 2}{0.1 \times 10^{-3}}$$

8.3.2 Diffraction Grating Equation

The diffraction grating equation describes the angular positions (θ) of diffraction maxima:

$$d\sin(\theta) = m\lambda$$

where d is the grating spacing, m is the order of the maximum, and λ is the wavelength.

Example: Diffraction Grating

For a diffraction grating with a spacing of 1×10^{-5} m and light of wavelength 600 nm, the angle of the first-order maximum can be calculated using the diffraction grating equation.

$$1 \times 10^{-5}\sin(\theta) = 1 \times 10^{-5} \times 600 \times 10^{-9}$$

8.3.3 Polarization

Polarization refers to the orientation of the electric field in a light wave. Malus's Law describes the intensity (I) of polarized light passed through an analyzer:

$$I = I_0 \cos^2(\theta)$$

where I_0 is the initial intensity and θ is the angle between the polarization direction and the analyzer.

Example: Malus's Law

If initially unpolarized light passes through an analyzer at an angle of 60°, the transmitted intensity can be calculated using Malus's Law.

$$I = I_0 \cos^2(60°)$$

8.3.4   Applications of Wave Optics

Wave optics finds applications in various technologies, including telecommunications and optical instruments.

Telecommunications

Understanding interference and diffraction is crucial in designing optical communication systems for efficient data transmission.

Optical Instruments

Wave optics principles are applied in the design of optical instruments such as microscopes and telescopes, enhancing their performance.

8.4   Sound Waves

Sound waves are mechanical waves that propagate through a medium, typically air. Understanding the properties and behavior of sound waves is essential in various applications, from music to communication.

8.4.1   Speed of Sound

The speed of sound (v_s) in a medium is given by the equation:

$$v_s = \sqrt{\frac{B}{\rho}}$$

where B is the bulk modulus of the medium and ρ is the density.

Example: Speed of Sound in Air

For air with a bulk modulus of $1.4 \times 10^5 \, \text{N/m}^2$ and a density of $1.2 \, \text{kg/m}^3$, the speed of sound can be calculated using the formula $v_s = \sqrt{\frac{B}{\rho}}$.

$$v_s = \sqrt{\frac{1.4 \times 10^5}{1.2}}$$

8.4.2 Wavelength and Frequency

The wavelength (λ) of a sound wave is related to its frequency (f) by the equation:

$$v_s = f\lambda$$

where v_s is the speed of sound.

Example: Wavelength and Frequency

If a sound wave with a frequency of $440 \, \text{Hz}$ travels in air with a speed of $340 \, \text{m/s}$, the wavelength can be calculated using $v_s = f\lambda$.

$$340 = 440\lambda$$

8.4.3 Intensity and Loudness

The intensity (I) of a sound wave is related to its amplitude (A) by the equation:

$$I = \frac{1}{2}\rho v_s f A^2$$

The loudness of a sound is often measured in decibels (dB):

$$\text{Loudness (dB)} = 10 \log_{10}\left(\frac{I}{I_0}\right)$$

where I_0 is the reference intensity.

Example: Sound Intensity

For a sound wave with amplitude 2×10^{-5} m and a frequency of 1000 Hz in air, the intensity can be calculated using $I = \frac{1}{2} \rho v_s f A^2$.

$$I = \frac{1}{2} \times 1.2 \times 340 \times 1000 \times (2 \times 10^{-5})^2$$

8.4.4   Doppler Effect

The Doppler effect describes the change in frequency or wavelength of a wave in relation to an observer moving relative to the source.

Example: Doppler Shift

If a car horn with a frequency of 500 Hz approaches a stationary observer at 20 m/s, the observed frequency can be calculated using the Doppler effect equation.

$$f' = f \left(\frac{v_s}{v_s - v_o} \right)$$

8.4.5   Applications of Sound Waves

Understanding sound waves is crucial in various applications, from music and communication to medical imaging.

Musical Instruments

Instruments produce sound waves based on principles of acoustics, contributing to the creation of music.

Communication Systems

Sound waves are used in communication systems, such as telephones and public address systems.

Medical Imaging

Ultrasound imaging utilizes sound waves for medical diagnostics without ionizing radiation.

Chapter 9

Electricity and Magnetism

9.1 Electric Charge and Fields

Electricity is a fundamental aspect of physics, and understanding electric charge and electric fields is crucial. This section explores the principles governing electric charge and the concept of electric fields.

9.1.1 Electric Charge

Electric charge is a fundamental property of matter, and it comes in two types: positive and negative. The unit of charge is the coulomb (C).

Coulomb's Law

Coulomb's Law describes the electrostatic force (F) between two point charges (q_1 and q_2) separated by a distance (r):

$$F = \frac{k \cdot |q_1 \cdot q_2|}{r^2}$$

where k is Coulomb's constant ($8.99 \times 10^9 \, \text{N m}^2/\text{C}^2$).

Example: Coulomb's Law

For two point charges, $q_1 = 5\,\text{nC}$ and $q_2 = -3\,\text{nC}$, separated by a distance of $2\,\text{m}$, the electrostatic force can be calculated using Coulomb's Law.

$$F = \frac{(8.99 \times 10^9) \cdot |5 \times 10^{-9} \cdot -3 \times 10^{-9}|}{(2)^2}$$

9.1.2   Electric Fields

The electric field (E) at a point in space is defined as the force (F) experienced by a positive test charge (q_0) placed at that point, divided by the magnitude of the test charge:

$$E = \frac{F}{q_0}$$

Electric Field Due to a Point Charge

The electric field (E) due to a point charge (Q) at a distance (r) is given by:

$$E = \frac{k \cdot |Q|}{r^2}$$

Example: Electric Field Due to a Point Charge

For a point charge $Q = 2\,\mu\text{C}$ located $3\,\text{m}$ away from a test charge $q_0 = 1\,\text{nC}$, the electric field can be calculated.

$$E = \frac{(8.99 \times 10^9) \cdot (2 \times 10^{-6})}{(3)^2}$$

9.1.3   Electric Potential Energy

The electric potential energy (U) of two point charges $(q_1$ and $q_2)$ separated by a distance (r) is given by:

$$U = \frac{k \cdot |q_1 \cdot q_2|}{r}$$

Example: Electric Potential Energy

For two point charges, $q_1 = 3\,\mu\text{C}$ and $q_2 = -4\,\mu\text{C}$, separated by a distance of 4 m, the electric potential energy can be calculated.

$$U = \frac{(8.99 \times 10^9) \cdot |3 \times 10^{-6} \cdot -4 \times 10^{-6}|}{4}$$

9.1.4 Applications of Electric Fields

Understanding electric charge and electric fields is fundamental to various technologies and applications.

Capacitors

Capacitors store electric charge and are used in electronic circuits for energy storage.

Electrostatic Precipitators

In environmental applications, electrostatic precipitators use electric fields to remove particulate matter from air.

Ionization in Gases

Electric fields play a crucial role in the ionization of gases, leading to phenomena like lightning.

9.2 Electric Circuits

Electric circuits are essential components of modern technology, enabling the flow of electric current and the functioning of various devices. This section explores key concepts related to electric circuits.

9.2.1   Ohm's Law

Ohm's Law describes the relationship between voltage (V), current (I), and resistance (R) in a circuit:

$$V = I \cdot R$$

Example: Ohm's Law

For a circuit with a voltage of $12\,\text{V}$ and a resistance of $4\,\Omega$, Ohm's Law can be used to calculate the current flowing through the circuit.

$$I = \frac{V}{R}$$

9.2.2   Series and Parallel Circuits

In series circuits, components are connected end-to-end, and the total resistance (R_{total}) is the sum of individual resistances. In parallel circuits, components are connected across the same voltage, and the reciprocal of the total resistance (R_{total}^{-1}) is the sum of reciprocals of individual resistances.

Example: Series and Parallel Circuits

Consider two resistors, $R_1 = 6\,\Omega$ and $R_2 = 4\,\Omega$, connected in series and parallel. The total resistance in each configuration can be calculated.

$$R_{\text{total, series}} = R_1 + R_2$$

$$R_{\text{total, parallel}}^{-1} = \frac{1}{R_1} + \frac{1}{R_2}$$

9.2.3   Kirchhoff's Laws

Kirchhoff's Laws are essential for analyzing complex circuits. Kirchhoff's Voltage Law (KVL) states that the sum of the voltages in any closed loop of a circuit is equal to the sum of the electromotive forces (emfs) in that loop. Kirchhoff's Current Law (KCL) states that the sum of currents entering a junction in a circuit is equal to the sum of currents leaving the junction.

Example: Kirchhoff's Laws

For a circuit with multiple resistors and voltage sources, applying KVL and KCL allows the determination of currents and voltages at different points in the circuit.

$$\sum V_{\text{loop}} = \sum \text{emf}_{\text{loop}}$$

$$\sum I_{\text{in}} = \sum I_{\text{out}}$$

9.2.4 RC Circuits

RC circuits involve the combination of resistors and capacitors. The time constant (τ) of an RC circuit is given by the product of resistance (R) and capacitance (C).

$$\tau = R \cdot C$$

Example: RC Circuit

For an RC circuit with $R = 2\,\Omega$ and $C = 3\,\text{F}$, the time constant can be calculated.

$$\tau = R \cdot C$$

9.2.5 Applications of Electric Circuits

Understanding electric circuits is crucial for various applications, including electronics, power distribution, and telecommunications.

Electronics

Electronic devices, such as computers and smartphones, rely on intricate circuits for their operation.

Power Distribution

Electric circuits play a vital role in the distribution of electrical power from power plants to homes and businesses.

Telecommunications

Circuits are integral to the functioning of telecommunication systems, facilitating the transmission of signals over long distances.

9.3   Magnetism

Magnetism is a fundamental force in nature, playing a crucial role in various technological applications. This section explores key concepts related to magnetism.

9.3.1   Magnetic Fields

A magnetic field (B) is the region around a magnet where magnetic forces act on other magnets or moving charged particles. The unit of magnetic field in the International System of Units (SI) is the tesla (T).

Magnetic Field Due to a Straight Current-Carrying Conductor

The magnetic field (B) at a distance (r) from a straight current-carrying conductor with current (I) is given by Ampère's Law:

$$B = \frac{\mu_0 \cdot I}{2\pi r}$$

where μ_0 is the permeability of free space ($4\pi \times 10^{-7}$ T m/A).

Example: Magnetic Field of a Conductor

For a straight conductor carrying a current of 5 A at a distance of 0.1 m, the magnetic field can be calculated.

$$B = \frac{(4\pi \times 10^{-7}) \cdot 5}{2\pi \cdot 0.1}$$

9.3.2 Magnetic Force on a Moving Charge

A charged particle moving in a magnetic field experiences a magnetic force (F_B) perpendicular to both its velocity (v) and the magnetic field (B). The magnitude of the magnetic force is given by:

$$F_B = |q \cdot v \cdot B|$$

where q is the charge of the particle.

Example: Magnetic Force on a Moving Charge

For an electron $(q = -1.6 \times 10^{-19}\,\text{C})$ moving with a velocity of $3 \times 10^7\,\text{m/s}$ in a magnetic field of 0.5 T, the magnetic force can be calculated.

$$F_B = |-1.6 \times 10^{-19} \cdot 3 \times 10^7 \cdot 0.5|$$

9.3.3 Magnetic Flux

Magnetic flux (Φ_B) is a measure of the magnetic field passing through a surface. It is given by the product of the magnetic field (B) and the perpendicular area (A) through which the field lines pass:

$$\Phi_B = B \cdot A$$

Faraday's Law of Electromagnetic Induction

Faraday's Law states that the electromotive force (EMF) induced in a coil is proportional to the rate of change of magnetic flux:

$$EMF = -N \cdot \frac{d\Phi_B}{dt}$$

where N is the number of turns in the coil.

Example: Magnetic Flux and Induced EMF

For a coil with $N = 100$ turns experiencing a change in magnetic flux of $2\,\text{T m}^2$ over $0.1\,\text{s}$, the induced EMF can be calculated.

$$EMF = -100 \cdot \frac{2}{0.1}$$

9.3.4   Applications of Magnetism

Understanding magnetism is crucial for various technological applications.

Magnetic Resonance Imaging (MRI)

MRI uses strong magnetic fields to create detailed images of internal body structures.

Electric Motors

Electric motors utilize magnetic fields to convert electrical energy into mechanical energy.

Magnetic Storage Devices

Magnetic fields are used in data storage devices like hard drives.

9.4 Electrostatics

Electrostatics deals with the study of electric charges and their interactions in a stationary state. This section explores key concepts related to electrostatics.

9.4.1 Coulomb's Law

Coulomb's Law describes the force (F) between two point charges (q_1 and q_2) separated by a distance (r):

$$F = k \cdot \frac{q_1 \cdot q_2}{r^2}$$

where k is Coulomb's constant ($8.99 \times 10^9 \, \text{N m}^2/\text{C}^2$).

Example: Coulomb's Law

For two point charges, $q_1 = 3 \times 10^{-6} \, \text{C}$ and $q_2 = -2 \times 10^{-6} \, \text{C}$, separated by a distance of $0.1 \, \text{m}$, the force between them can be calculated.

$$F = (8.99 \times 10^9) \cdot \frac{3 \times 10^{-6} \cdot (-2 \times 10^{-6})}{(0.1)^2}$$

9.4.2 Electric Field

The electric field (E) at a point in space is the force per unit positive charge that a test charge would experience at that point. It is given by:

$$E = \frac{F}{q_0}$$

where q_0 is the test charge.

Example: Electric Field

For a point charge $Q = 2 \times 10^{-6} \, \text{C}$ creating an electric field at a distance of $0.5 \, \text{m}$, the electric field can be calculated.

$$E = \frac{F}{q_0}$$

9.4.3   Gauss's Law

Gauss's Law relates the electric flux (Φ_E) through a closed surface to the total charge enclosed (Q_{enc}):

$$\Phi_E = \frac{Q_{\text{enc}}}{\varepsilon_0}$$

where ε_0 is the permittivity of free space ($8.85 \times 10^{-12}\,\text{C}^2/\text{N m}^2$).

Example: Gauss's Law

For a spherical surface containing a total charge of $Q_{\text{enc}} = 1 \times 10^{-5}\,\text{C}$, the electric flux through the surface can be calculated.

$$\Phi_E = \frac{Q_{\text{enc}}}{\varepsilon_0}$$

9.4.4   Electric Potential

The electric potential (V) at a point in space is the electric potential energy per unit charge. It is given by:

$$V = \frac{U}{q_0}$$

where U is the electric potential energy.

Example: Electric Potential

For a point charge $Q = -4 \times 10^{-6}\,\text{C}$ at a distance of $0.2\,\text{m}$, the electric potential at that point can be calculated.

$$V = \frac{U}{q_0}$$

9.4.5   Applications of Electrostatics

Understanding electrostatics is crucial for various technological applications.

Capacitors

Capacitors store electric charge and are widely used in electronic circuits.

Electrostatic Precipitators

Electrostatic precipitators remove particulate matter from industrial exhaust gases.

Inkjet Printers

Inkjet printers use electrostatic forces to propel ink droplets onto paper.

Chapter 10

Modern Physics

10.1 Special Relativity

Special relativity, developed by Albert Einstein, revolutionized our understanding of space and time. This section explores the fundamental concepts of special relativity.

10.1.1 Postulates of Special Relativity

Special relativity is based on two postulates:

1. The laws of physics are the same for all observers in uniform motion (inertial frames of reference).

2. The speed of light (c) is the same for all observers, regardless of their motion or the motion of the light source.

10.1.2 Lorentz Transformation

The Lorentz transformation equations relate the coordinates (x, y, z, t) of an event in one inertial frame (S) to the coordinates (x', y', z', t') in another inertial frame (S') moving with a constant velocity v relative to S:

$$x' = \gamma(x - vt)$$
$$y' = y$$
$$z' = z$$
$$t' = \gamma\left(t - \frac{vx}{c^2}\right)$$

where $\gamma = \frac{1}{\sqrt{1 - \frac{v^2}{c^2}}}$.

Example: Lorentz Transformation

For an event with coordinates $(x, t) = (3\,\text{m}, 1\,\text{s})$ in frame S, moving with $v = 0.8c$ relative to frame S', the transformed coordinates in S' can be calculated.

$$x' = \gamma(x - vt)$$
$$t' = \gamma\left(t - \frac{vx}{c^2}\right)$$

10.1.3   Time Dilation

Time dilation occurs when an observer in one inertial frame measures the time interval (Δt) between two events while another observer in a different inertial frame measures a longer time interval $(\Delta t')$. The relationship between these time intervals is given by:

$$\Delta t' = \frac{\Delta t}{\gamma}$$

Example: Time Dilation

For a moving clock with $\Delta t = 2\,\text{s}$ in frame S, moving with $v = 0.9c$ relative to frame S', the observed time interval $\Delta t'$ can be calculated.

$$\Delta t' = \frac{\Delta t}{\gamma}$$

10.1.4 Length Contraction

Length contraction occurs when an observer in one frame measures the length
(L) of an object moving with velocity v in the direction of its motion. The
contracted length (L') is given by:

$$L' = \frac{L}{\gamma}$$

Example: Length Contraction

For a rod with $L = 10\,\text{m}$ at rest in frame S, the contracted length L' can be
calculated for a frame S' moving with $v = 0.95c$.

$$L' = \frac{L}{\gamma}$$

10.1.5 Relativistic Energy and Mass

The total energy (E) of an object in motion is related to its rest mass (m_0) and
velocity (v) by the equation:

$$E = \gamma m_0 c^2$$

where c is the speed of light.

Example: Relativistic Energy

For a particle with rest mass $m_0 = 1\,\text{kg}$ moving at $v = 0.99c$, the relativistic
energy E can be calculated.

$$E = \gamma m_0 c^2$$

10.2 Quantum Mechanics

Quantum mechanics revolutionized our understanding of the microscopic world,
describing the behavior of particles at the quantum level. This section explores

fundamental concepts in quantum mechanics.

10.2.1　Wave-Particle Duality

Wave-particle duality is a key concept in quantum mechanics, suggesting that particles, such as electrons, exhibit both wave and particle-like properties. The de Broglie wavelength (λ) is given by:

$$\lambda = \frac{h}{p}$$

where h is Planck's constant and p is the momentum of the particle.

Example: de Broglie Wavelength

For an electron with momentum $p = 1 \times 10^{-24}$ kg m/s, the de Broglie wavelength can be calculated.

$$\lambda = \frac{h}{p}$$

10.2.2　Quantum Operators and Wavefunctions

In quantum mechanics, physical observables are represented by operators. The wavefunction (Ψ) describes the state of a quantum system and is governed by the Schrödinger equation:

$$\hat{H}\Psi = E\Psi$$

where $\hat{H}$ is the Hamiltonian operator and E is the energy of the system.

Example: Schrödinger Equation

For a particle in a one-dimensional box, the Schrödinger equation can be solved to find the allowed energy levels and corresponding wavefunctions.

$$\hat{H}\Psi = E\Psi$$

10.2.3 Heisenberg Uncertainty Principle

The Heisenberg Uncertainty Principle states that it is impossible to simultaneously know the exact position (Δx) and momentum (Δp) of a particle with absolute certainty:

$$\Delta x \cdot \Delta p \geq \frac{\hbar}{2}$$

where $\hbar$ is the reduced Planck's constant.

Example: Uncertainty Principle

For a particle with uncertainty in position $\Delta x = 1\,\text{nm}$, the minimum uncertainty in momentum Δp can be calculated.

$$\Delta x \cdot \Delta p \geq \frac{\hbar}{2}$$

10.2.4 Quantum Tunneling

Quantum tunneling is the phenomenon where particles can pass through a potential barrier that classical mechanics predicts they cannot overcome. The probability of tunneling (T) is given by:

$$T = e^{-2ak}$$

where a is the width of the barrier and k is the wave vector.

Example: Quantum Tunneling

For a particle encountering a potential barrier with $a = 1\,\text{nm}$ and $k = 2 \times 10^{10}\,\text{m}^{-1}$, the probability of tunneling can be calculated.

$$T = e^{-2ak}$$

10.2.5   Applications of Quantum Mechanics

Quantum mechanics has led to numerous technological advancements, including the development of semiconductors, lasers, and quantum computers.

Semiconductors

The behavior of electrons in semiconductors is described by quantum mechanics, forming the basis for electronic devices.

Lasers

Lasers operate based on quantum principles, involving the emission of photons due to transitions between energy levels.

Quantum Computers

Quantum computers leverage quantum bits (qubits) and quantum entanglement for exponentially faster computation.

10.3   Particle Physics

Particle physics explores the fundamental constituents of matter and their interactions at the smallest scales. This section delves into the key concepts in particle physics.

10.3.1   Elementary Particles

Elementary particles are the building blocks of matter. The Standard Model classifies elementary particles into fermions and bosons.

Fermions

Fermions are particles with half-integer spin, including quarks and leptons. The six types of quarks are up, down, charm, strange, top, and bottom. Leptons include electrons, muons, and tau particles.

Bosons

Bosons are particles with integer spin, responsible for mediating forces. The photon mediates electromagnetism, while the W and Z bosons mediate the weak force. The gluon mediates the strong force, and the recently discovered Higgs boson provides mass to other particles.

10.3.2 Quantum Chromodynamics (QCD)

Quantum Chromodynamics describes the strong force between quarks mediated by gluons. The theory explains how quarks and gluons interact within hadrons.

Example: QCD and Gluons

For a quark-antiquark pair, QCD predicts the exchange of gluons. The color charge of quarks and gluons is a fundamental aspect of QCD.

10.3.3 Electroweak Theory

The electroweak theory unifies electromagnetism and the weak nuclear force into a single theoretical framework. The symmetry-breaking mechanism involves the Higgs field and results in the masses of W and Z bosons.

Example: Electroweak Unification

The electroweak theory predicts the existence of neutral currents and charged currents. The unified theory successfully describes a range of experimental results.

10.3.4 Collider Experiments

Particle physicists study elementary particles by colliding them at high energies. Large colliders, such as the Large Hadron Collider (LHC), provide insights into the fundamental forces and particles.

Example: LHC and Higgs Boson Discovery

The discovery of the Higgs boson at the LHC in 2012 confirmed the existence of the last missing piece of the Standard Model.

10.3.5   Beyond the Standard Model

While the Standard Model has been remarkably successful, there are phenomena it does not explain, such as dark matter and dark energy. The search for physics beyond the Standard Model continues.

Example: Dark Matter Candidates

Various theoretical candidates for dark matter include weakly interacting massive particles (WIMPs) and axions. Experiments aim to detect these elusive particles.

10.3.6   Applications of Particle Physics

Particle physics has led to technological advancements and practical applications, including medical imaging technologies and particle therapy for cancer treatment.

Medical Imaging

Techniques developed in particle physics, such as positron emission tomography (PET) and magnetic resonance imaging (MRI), have revolutionized medical diagnostics.

Particle Therapy

Particle beams, such as protons and carbon ions, are used for precise cancer treatment, minimizing damage to surrounding healthy tissues.

Chapter 11

Optics

11.1 Wave Nature of Light

The wave nature of light is a fundamental aspect of optics, describing the behavior of light waves. This section explores key concepts related to the wave nature of light.

11.1.1 Wave Characteristics

Light exhibits wave-like properties, including wavelength (λ), frequency (f), and wave speed (v). The relationship between these parameters is given by the wave equation:

$$v = f\lambda$$

where v is the speed of light.

Example: Wave Characteristics

For light with a wavelength of $\lambda = 500\,\text{nm}$ in a vacuum, calculate its frequency and wave speed.

105

$$v = f\lambda$$

11.1.2   Interference and Diffraction

Interference and diffraction are phenomena that occur when light waves interact. Interference involves the superposition of waves, leading to constructive or destructive interference. Diffraction is the bending of light waves around obstacles.

Example: Interference

Two coherent light sources with a path difference of $\Delta d = \lambda/2$ interfere. Determine the conditions for constructive and destructive interference.

$$\Delta d = m\lambda$$

Example: Diffraction

A single slit of width a is illuminated with monochromatic light. Calculate the angular positions of the first-order diffraction minima.

$$a\sin\theta = m\lambda$$

11.1.3   Polarization of Light

Polarization refers to the orientation of light waves' oscillations. Polarizers selectively transmit light waves with a specific polarization direction.

Example: Polarization

Linearly polarized light passes through a polarizer with an angle θ between the polarization direction and the polarizer axis. Determine the intensity of the transmitted light.

$$I = I_0 \cos^2 \theta$$

11.1.4 Huygens' Principle

Huygens' Principle states that each point on a wavefront can be considered as a source of secondary spherical waves. The envelope of these secondary waves forms the new wavefront.

Example: Huygens' Principle

Apply Huygens' Principle to explain the formation of a circular wavefront when light passes through a small aperture.

$$\text{Wavefront at time } t = \text{Wavefront at time } t - \frac{d}{v}$$

11.1.5 Applications of Wave Optics

Wave optics plays a crucial role in various optical devices, including diffraction gratings, interference filters, and holography.

Diffraction Gratings

Diffraction gratings utilize the interference pattern produced by multiple slits to disperse light into its component wavelengths.

Interference Filters

Interference filters selectively transmit or reflect certain wavelengths of light based on interference effects.

Holography

Holography involves recording and reconstructing three-dimensional images using the wave nature of light.

11.2   Interference and Diffraction

Interference and diffraction are phenomena that occur when light waves interact. Understanding these phenomena is crucial in optics, providing insights into wave behavior.

11.2.1   Interference

Interference is the superposition of two or more waves, leading to the reinforcement (constructive interference) or cancellation (destructive interference) of amplitudes.

Coherent Sources

For interference to occur, sources must be coherent, meaning they have a constant phase difference. Two sources with a path difference Δd exhibit constructive interference when $\Delta d = m\lambda$, and destructive interference when $\Delta d = (m + 0.5)\lambda$.

Example: Double-Slit Interference

Consider two coherent light sources producing waves that interfere on a screen. If the path difference between the sources is $\Delta d = 2\,\text{mm}$, and the wavelength of light is $\lambda = 600\,\text{nm}$, determine the interference pattern on the screen.

$$\Delta d = m\lambda$$

11.2.2   Diffraction

Diffraction is the bending of waves around obstacles or the spreading of waves when they encounter an aperture. It results in the formation of interference patterns.

Single-Slit Diffraction

For a single slit of width a, the angular positions of diffraction minima are given by $a \sin \theta = m\lambda$.

Example: Single-Slit Diffraction

A laser beam with a wavelength $\lambda = 700\,\text{nm}$ passes through a single slit of width $a = 0.1\,\text{mm}$. Determine the angular positions of the first-order diffraction minima.

$$a \sin \theta = m\lambda$$

11.2.3 Applications

Interference and diffraction find applications in various optical devices, including diffraction gratings, interference filters, and spectrometers.

Diffraction Gratings

Diffraction gratings utilize the interference pattern produced by multiple slits to disperse light into its component wavelengths.

Interference Filters

Interference filters selectively transmit or reflect certain wavelengths of light based on interference effects.

Spectrometers

Spectrometers use interference and diffraction principles to analyze the spectrum of light emitted or absorbed by a substance.

11.3   Polarization

Polarization refers to the orientation of the oscillations of light waves. Understanding polarization is essential in optics, particularly in the design of optical devices.

11.3.1   Polarization States

Light waves can be linearly, circularly, or elliptically polarized. Linear polarization occurs when the electric field oscillates in a straight line.

Linear Polarization

For linearly polarized light, the intensity transmitted through a polarizer with an angle θ is given by $I = I_0 \cos^2 \theta$, where I_0 is the initial intensity.

Example: Linear Polarization

Linearly polarized light passes through a polarizer with an angle $\theta = 45°$. Determine the intensity of the transmitted light if the initial intensity is $I_0 = 10\,\text{W/m}^2$.

$$I = I_0 \cos^2 \theta$$

11.3.2   Malus's Law

Malus's Law describes the intensity of light transmitted through a polarizer. It states that the transmitted intensity (I) is proportional to the square of the cosine of the angle (θ) between the polarizer axis and the light's polarization direction.

Circular Polarization

For circularly polarized light, the electric field vector traces a circle as the wave propagates.

Elliptical Polarization

Elliptically polarized light results from a combination of linear and circular polarization.

11.3.3 Applications of Polarization

Polarization has practical applications in various optical devices, including sunglasses, liquid crystal displays (LCDs), and 3D glasses.

Sunglasses

Polarized sunglasses reduce glare by selectively blocking horizontally polarized light.

LCDs

LCDs use polarizers to control the transmission of light and produce images with adjustable brightness.

3D Glasses

3D glasses use polarization to separate images intended for the left and right eyes.

Chapter 12

Electromagnetic Waves

12.1 Maxwell's Equations

Maxwell's equations are a set of four fundamental equations that describe the behavior of electric and magnetic fields in space and time. They form the basis for understanding the generation and propagation of electromagnetic waves.

12.1.1 Gauss's Law for Electricity

The first equation, Gauss's Law for Electricity, relates the electric flux through a closed surface (Φ_E) to the enclosed charge (Q_{enc}).

$$\oint \vec{E} \cdot d\vec{A} = \frac{Q_{\text{enc}}}{\varepsilon_0}$$

Example: Electric Flux

Calculate the electric flux through a closed surface with an enclosed charge of $Q_{\text{enc}} = 2\,\text{nC}$ and a permittivity of free space (ε_0) of $8.85 \times 10^{-12}\,\text{C}^2/\text{N} \cdot \text{m}^2$.

$$\oint \vec{E} \cdot d\vec{A} = \frac{Q_{\text{enc}}}{\varepsilon_0}$$

113

12.1.2   Gauss's Law for Magnetism

The second equation, Gauss's Law for Magnetism, states that the magnetic flux through any closed surface is zero.

$$\oint \vec{B} \cdot d\vec{A} = 0$$

Magnetic Monopoles

Gauss's Law for Magnetism suggests the absence of magnetic monopoles.

12.1.3   Faraday's Law of Induction

The third equation, Faraday's Law of Induction, describes the electromotive force ($\mathcal{E}$) induced in a closed loop due to a changing magnetic flux (Φ_B).

$$\mathcal{E} = -\frac{d\Phi_B}{dt}$$

Example: Induced EMF

A coil with 100 turns experiences a magnetic flux change of $5\,\mathrm{mT} \cdot \mathrm{m}^2$ over 2 seconds. Calculate the induced electromotive force.

$$\mathcal{E} = -\frac{d\Phi_B}{dt}$$

12.1.4   Ampère's Law with Maxwell's Addition

The fourth equation, Ampère's Law with Maxwell's Addition, relates the magnetic field ($\vec{B}$) to the current density ($\vec{J}$) and the rate of change of the electric field ($\vec{E}$).

$$\oint \vec{B} \cdot d\vec{l} = \mu_0 \left(\int \vec{J} \cdot d\vec{A} + \varepsilon_0 \frac{d}{dt} \int \vec{E} \cdot d\vec{A} \right)$$

Maxwell's Addition

Maxwell's addition introduces the displacement current term, accounting for time-varying electric fields.

12.1.5 Significance and Applications

Maxwell's equations are foundational in understanding electromagnetic phenomena, from the behavior of electric and magnetic fields to the generation and propagation of electromagnetic waves, including radio waves, microwaves, and light.

Electromagnetic Waves

The combined set of Maxwell's equations predicts the existence of electromagnetic waves, propagating at the speed of light.

Technological Impact

Maxwell's equations have had a profound impact on technology, enabling the development of devices such as antennas, radar systems, and communication systems.

12.2 Electromagnetic Spectrum

The electromagnetic spectrum encompasses a broad range of electromagnetic waves, each with its unique properties and applications. It is divided into different regions based on wavelength and frequency.

12.2.1 Overview of the Electromagnetic Spectrum

The electromagnetic spectrum includes radio waves, microwaves, infrared radiation, visible light, ultraviolet radiation, X-rays, and gamma rays. Each region has specific characteristics and interactions with matter.

12.2.2   Formulas Relevant to Electromagnetic Waves

Speed of Light

The speed of light (c) is a fundamental constant related to the wavelength (λ) and frequency (f) of electromagnetic waves.

$$c = \lambda f$$

Energy of Photons

The energy (E) of a photon is directly proportional to its frequency (f).

$$E = hf$$

Where h is Planck's constant (6.626×10^{-34} J $\cdot$ s).

12.2.3   Radio Waves

Radio waves have the longest wavelength in the electromagnetic spectrum, ranging from about a millimeter to hundreds of meters. They are used for communication, including broadcasting and mobile phones.

Example: Broadcasting

Discuss the use of radio waves in broadcasting and how antennas transmit and receive signals.

12.2.4   Microwaves

Microwaves have shorter wavelengths than radio waves and are commonly used for cooking and communication, especially in microwave ovens and satellite communication.

Example: Microwave Oven

Explain how microwave ovens work, focusing on the interaction between microwaves and water molecules in food.

12.2.5 Infrared Radiation

Infrared radiation has longer wavelengths than visible light and is known for its heat-emitting properties. Infrared is used in various applications, including thermal imaging and remote sensing.

Example: Infrared Cameras

Explore how infrared cameras capture thermal images and their applications in industries such as healthcare and security.

12.2.6 Visible Light

Visible light is the range of electromagnetic waves that our eyes can detect. It is essential for vision and has various applications in lighting and imaging.

Example: Optics

Discuss the principles of optics and how visible light is refracted and reflected, leading to applications in lenses and mirrors.

12.2.7 Ultraviolet Radiation

Ultraviolet (UV) radiation has shorter wavelengths than visible light and is known for its effects on skin and materials. UV is used in medical applications and sterilization.

Example: UV Sterilization

Examine how UV radiation is used for sterilizing surfaces and air, particularly in healthcare settings.

12.2.8   X-rays and Gamma Rays

X-rays and gamma rays have the shortest wavelengths and high energy. They are used in medical imaging, cancer treatment, and materials testing.

Example: Medical Imaging

Explore how X-rays are used in medical imaging, emphasizing their ability to penetrate tissues for diagnostic purposes.

Chapter 13

Nuclear Physics

13.1 Radioactivity

Radioactivity is a fundamental phenomenon in nuclear physics where unstable atomic nuclei undergo spontaneous decay, emitting radiation in the form of alpha (α, helium nuclei), beta (β, electrons or positrons), or gamma (γ, electromagnetic) particles. This section explores the principles of radioactivity, relevant formulas, and applications.

13.1.1 Fundamental Concepts

Decay Law

The decay of radioactive substances follows an exponential decay law:

$$N(t) = N_0 e^{-\lambda t}$$

where $N(t)$ is the remaining quantity at time t, N_0 is the initial quantity, λ is the decay constant, and e is the base of the natural logarithm.

Half-Life

The half-life $(T_{1/2})$ is the time required for half of the radioactive substance to decay. It is related to the decay constant:

$$T_{1/2} = \frac{\ln(2)}{\lambda}$$

13.1.2   Alpha Decay

Alpha decay involves the emission of an alpha particle (α), consisting of two protons and two neutrons.

Example: Uranium Decay

Discuss the alpha decay of uranium-238 into thorium-234, including the resulting daughter nucleus and emitted alpha particle.

13.1.3   Beta Decay

Beta decay occurs when a neutron transforms into a proton with the emission of a beta particle (β^-) or when a proton transforms into a neutron with the emission of a positron (β^+).

Example: Carbon-14 Dating

Explain the use of carbon-14 (^{14}C) in radiocarbon dating, emphasizing the beta decay of carbon-14 to nitrogen-14.

13.1.4   Gamma Decay

Gamma decay involves the emission of a gamma ray (γ), which is a high-energy electromagnetic wave.

Example: Medical Imaging

Explore the use of gamma rays in medical imaging, such as positron emission tomography (PET) scans.

13.1.5 Nuclear Reactions

Nuclear reactions involve changes in atomic nuclei, including fusion and fission.

Example: Nuclear Power

Discuss nuclear fission as a source of energy in nuclear power plants, highlighting the reaction of uranium-235.

13.1.6 Radiation Protection

Understand the principles of radiation protection, including the use of shielding and monitoring devices.

Example: Radiography

Examine safety measures in industrial radiography, where X-rays or gamma rays are used for material testing.

13.2 Nuclear Reactions

Nuclear reactions involve changes in atomic nuclei, including both fusion and fission processes. Understanding these reactions is crucial for various applications, ranging from energy production in nuclear power plants to the study of cosmic processes.

13.2.1 Fission Reactions

Nuclear fission is the process where a heavy nucleus splits into two lighter nuclei, releasing a large amount of energy. The reaction can be represented as:

$$\,^A_Z X +\,^1_0 n \rightarrow\,^{A_1}_{Z_1} Y_1 +\,^{A_2}_{Z_2} Y_2 + \text{energy}$$

where $^A_Z X$ is the original nucleus, $^1_0 n$ is a neutron, and $^{A_1}_{Z_1} Y_1$ and $^{A_2}_{Z_2} Y_2$ are the resulting fragments.

Example: Uranium-235 Fission

Describe the fission of uranium-235, including the products and energy released.

13.2.2   Fusion Reactions

Nuclear fusion involves combining two light nuclei to form a heavier nucleus, releasing a significant amount of energy. The reaction can be represented as:

$$\,^{A_1}_{Z_1} X_1 +\,^{A_2}_{Z_2} X_2 \rightarrow\,^A_Z Y +\,^1_0 n + \text{energy}$$

where $^{A_1}_{Z_1} X_1$ and $^{A_2}_{Z_2} X_2$ are the initial nuclei, $^A_Z Y$ is the resulting nucleus, and $^1_0 n$ is a neutron.

Example: Deuterium-Tritium Fusion

Explain the fusion of deuterium and tritium, emphasizing its potential application in controlled thermonuclear reactions.

13.2.3   Cross Section and Reaction Rate

The cross section (σ) represents the effective area for a nuclear reaction, and the reaction rate (R) is the number of reactions per unit time:

$$R = n \cdot \sigma \cdot v$$

where n is the number density of target nuclei and v is the relative velocity of the interacting particles.

Example: Neutron Activation

Discuss the concept of neutron activation, where materials exposed to neutrons become radioactive.

13.2.4   Nuclear Transmutation

Nuclear transmutation involves changing one element into another through nuclear reactions.

Example: Isotope Production

Explore the production of medical isotopes through nuclear transmutation, such as technetium-99m for diagnostic imaging.

13.2.5   Applications of Nuclear Reactions

Example: Nuclear Power Plants

Examine the use of nuclear reactions for electricity generation in nuclear power plants, emphasizing the controlled fission reactions.

13.3   Elementary Particle Physics

Elementary Particle Physics explores the fundamental constituents of matter and their interactions. The Standard Model is the theoretical framework that describes the known elementary particles and their electromagnetic, weak, and strong interactions.

13.3.1   Fundamental Particles

The basic building blocks of matter are quarks and leptons. Quarks combine to form hadrons, while leptons do not experience strong interactions.

Quarks

Quarks come in six flavors: up (u, c, t) and down (d, s, b). They carry fractional electric charges.

Leptons

Leptons include the electron (e), muon (μ), tau (τ), and their associated neutrinos (ν_e, ν_μ, ν_τ).

13.3.2   Gauge Bosons

Gauge bosons mediate the fundamental forces:

Photon (γ)

The photon mediates electromagnetic interactions between charged particles.

W and Z Bosons

The W and Z bosons mediate weak interactions, responsible for processes like beta decay.

Gluon

Gluons mediate the strong force between quarks, binding them into hadrons.

13.3.3   Higgs Boson

The Higgs boson gives mass to elementary particles through the Higgs mechanism.

13.3.4   Feynman Diagrams

Feynman diagrams visually represent particle interactions in quantum field theory.

Example: Electron-Positron Annihilation

Describe the Feynman diagram for electron-positron annihilation, resulting in the creation of photons.

13.3.5 Experimental Techniques

Particle accelerators and detectors are crucial tools for studying elementary particles.

Example: Large Hadron Collider (LHC)

Explain the role of the LHC in discovering the Higgs boson and exploring high-energy particle physics.

13.3.6 Beyond the Standard Model

Discuss challenges and open questions in particle physics, such as dark matter, dark energy, and the quest for a unified theory.

Example: Neutrino Oscillations

Explore neutrino oscillations as experimental evidence beyond the Standard Model.

13.3.7 Applications

Example: Medical Imaging

Highlight the application of particle physics technologies in medical imaging, such as positron emission tomography (PET).

Chapter 14

Astrophysics

14.1   Introduction to Astrophysics

Astrophysics is the branch of physics that studies celestial objects and phenomena beyond Earth's atmosphere. It combines principles from astronomy and physics to understand the nature of the universe.

14.1.1   Key Concepts

Celestial Objects

Celestial objects include stars, planets, galaxies, and cosmic structures. Each plays a unique role in the cosmic ballet.

Cosmology

Cosmology explores the large-scale structure and evolution of the universe, addressing questions about its origin and ultimate fate.

Gravity and Celestial Mechanics

Newton's law of gravity and Kepler's laws of planetary motion are foundational to understanding the motion of celestial bodies.

14.1.2   Formulas

Newton's Law of Gravity

The gravitational force (F) between two masses (m_1 and m_2) separated by a distance (r) is given by:

$$F = G\frac{m_1 \cdot m_2}{r^2}$$

where G is the gravitational constant.

Kepler's Third Law

The square of the orbital period (T) of a celestial body is proportional to the cube of its average distance (a) from the central body:

$$T^2 \propto a^3$$

14.1.3   Working Examples

Example: Orbits of Planets

Calculate the orbital period of a hypothetical planet at a given distance from its star using Kepler's third law.

Example: Gravitational Force

Determine the gravitational force between two stars given their masses and separation distance.

14.1.4   Astrophysical Phenomena

Stellar Nucleosynthesis

Explore the process by which stars fuse elements in their cores, releasing energy and creating heavier elements.

Black Holes

Understand the concept of black holes, regions of spacetime exhibiting gravitational forces so strong that nothing can escape them.

14.1.5 Observational Techniques

Telescopes

Discuss the role of telescopes in observing celestial objects and how different types of telescopes capture various wavelengths of light.

Spectroscopy

Explain how spectroscopy helps astronomers analyze the composition and temperature of celestial bodies.

14.2 Stellar Astrophysics

Stellar astrophysics is a branch of astrophysics that studies the properties, structure, and evolution of stars. It involves understanding the physical processes occurring within stars and how they influence the universe.

14.2.1 Key Concepts

Stellar Structure

Stars exhibit layers of varying composition and temperature, from the core where nuclear fusion occurs to the outer layers where radiation escapes into space.

Stellar Evolution

Stars undergo various stages of evolution, from their formation through to their eventual fate, which is influenced by their mass.

Stellar Spectra

The study of stellar spectra provides valuable information about a star's temperature, composition, and motion.

14.2.2   Formulas

Stellar Luminosity

The luminosity (L) of a star is related to its radius (R) and effective temperature (T_{eff}) by the Stefan-Boltzmann law:

$$L = 4\pi R^2 \sigma T_{\text{eff}}^4$$

where σ is the Stefan-Boltzmann constant.

Mass-Luminosity Relation

For main-sequence stars, the mass-luminosity relation is often expressed as:

$$\frac{L}{L_\odot} = \left(\frac{M}{M_\odot}\right)^\alpha$$

where $L_\odot$ and $M_\odot$ are the solar luminosity and mass, respectively.

14.2.3   Working Examples

Example: Determining Stellar Luminosity

Calculate the luminosity of a main-sequence star with a given radius and effective temperature.

Example: Mass-Luminosity Relation

Explore how the mass-luminosity relation helps predict the luminosity of stars based on their masses.

14.2.4   Stellar Types

Main-Sequence Stars

Discuss the characteristics and properties of main-sequence stars, which represent the majority of stars in the universe.

Red Giants and White Dwarfs

Examine the evolutionary stages of stars like red giants and white dwarfs, highlighting their unique features.

14.2.5   Observations and Discoveries

Hertzsprung-Russell (HR) Diagram

Introduce the HR diagram, a valuable tool for classifying stars based on their luminosity and temperature.

Nuclear Fusion in Stars

Explain how nuclear fusion processes in stellar cores lead to the release of energy and the formation of heavier elements.

14.3   Cosmology

Cosmology is the study of the large-scale structure, origin, evolution, and eventual fate of the universe. It involves understanding the fundamental properties of space, time, and the contents of the universe.

14.3.1   Key Concepts

Big Bang Theory

The Big Bang theory is the prevailing cosmological model that describes the origin of the universe as a hot, dense state from which it expanded and cooled over time.

Dark Matter and Dark Energy

Cosmologists investigate the mysterious components of the universe known as dark matter and dark energy, which collectively constitute the majority of the cosmic content.

Cosmic Microwave Background (CMB)

The CMB is the residual radiation from the Big Bang and provides valuable insights into the early universe's conditions.

14.3.2 Formulas

Hubble's Law

Hubble's law describes the relationship between the recessional velocity (v) of galaxies and their distance (d):

$$v = H_0 \cdot d$$

where H_0 is the Hubble constant.

Friedmann Equations

The Friedmann equations govern the expansion of the universe and are expressed in terms of the scale factor (a):

$$\left(\frac{\dot{a}}{a}\right)^2 = \frac{8\pi G}{3}\rho - \frac{k}{a^2}$$

where ρ is the energy density and k represents the spatial curvature.

14.3.3 Working Examples

Example: Calculating the Age of the Universe

Use Hubble's law to estimate the age of the universe based on the observed recessional velocities of galaxies.

Example: Understanding Dark Matter's Influence

Explore the impact of dark matter on the large-scale structure of the universe using simulations and observational data.

14.3.4 Observations and Discoveries

Expansion of the Universe

Discuss the historical observations that led to the realization that the universe is expanding.

Cosmic Inflation

Introduce the concept of cosmic inflation, a brief period of exponential expansion thought to have occurred shortly after the Big Bang.

14.3.5 Cosmic Timeline

Early Universe

Examine the conditions and events in the universe's early moments, including nucleosynthesis and the formation of the first atoms.

Formation of Large-Scale Structures

Trace the evolution of the cosmic web, from the initial density fluctuations to the formation of galaxies and galaxy clusters.

Chapter 15

Fluid Mechanics

15.1 Fluid Properties and Fluid Statics

Fluid mechanics is the study of fluids—liquids and gases—and their properties, behavior, and motion. This section focuses on fundamental concepts related to fluid properties and fluid statics.

15.1.1 Fluid Properties

Density (ρ)

Density is the mass per unit volume and is a fundamental property of fluids.

$$\rho = \frac{m}{V}$$

where m is the mass and V is the volume.

Pressure (P)

Pressure is defined as force per unit area and is a crucial property in fluid mechanics.

$$P = \frac{F}{A}$$

where F is the force applied perpendicular to the surface and A is the area.

Viscosity (η)

Viscosity measures a fluid's resistance to flow. It is the internal friction within the fluid.

15.1.2   Fluid Statics

Pascal's Law

Pascal's Law states that any change in pressure applied at any point in an enclosed fluid is transmitted undiminished throughout the fluid.

$$P_1 = P_2$$

where P_1 and P_2 are pressures at different points.

Hydrostatic Pressure

The pressure exerted by a fluid at rest is given by the hydrostatic pressure formula:

$$P = P_0 + \rho g h$$

where P_0 is the atmospheric pressure, ρ is the density of the fluid, g is the acceleration due to gravity, and h is the depth.

15.1.3   Formulas

Buoyancy Force

The buoyancy force (F_b) acting on an object submerged in a fluid is given by Archimedes' principle:

$$F_b = \rho_{\text{fluid}} \cdot V_{\text{displaced}} \cdot g$$

where ρ_{fluid} is the density of the fluid, $V_{\text{displaced}}$ is the volume of fluid displaced, and g is the acceleration due to gravity.

Manometer Equation

The manometer equation relates the pressure difference in a U-tube manometer to the height difference of the fluid columns:

$$P_A - P_B = \rho g h$$

where P_A and P_B are pressures at different points, ρ is the fluid density, g is the acceleration due to gravity, and h is the height difference.

15.1.4 Working Examples

Example: Buoyant Force

Calculate the buoyant force acting on a fully submerged object with a given volume and fluid density.

Example: Hydrostatic Pressure

Determine the hydrostatic pressure at different depths in a fluid, considering the atmospheric pressure and fluid density.

15.2 Fluid Dynamics

Fluid dynamics explores the motion of fluids—liquids and gases. This section delves into the fundamental principles and equations governing fluid flow, providing a basis for understanding the behavior of fluids in motion.

15.2.1 Continuity Equation

The continuity equation expresses the conservation of mass for an incompressible fluid. For steady flow through a pipe, it is given by:

$$A_1 v_1 = A_2 v_2$$

where A is the cross-sectional area and v is the velocity at different points.

15.2.2   Bernoulli's Equation

Bernoulli's equation describes the relationship between pressure, velocity, and elevation in a fluid flow:

$$P + \frac{1}{2}\rho v^2 + \rho gh = \text{constant}$$

where P is pressure, ρ is density, v is velocity, g is acceleration due to gravity, and h is height.

15.2.3   Navier-Stokes Equations

The Navier-Stokes equations are a set of differential equations describing fluid motion. For an incompressible, Newtonian fluid, they are given by:

$$\rho \left(\frac{\partial \mathbf{v}}{\partial t} + \mathbf{v} \cdot \nabla \mathbf{v} \right) = -\nabla P + \mu \nabla^2 \mathbf{v} + \rho \mathbf{g}$$

where ρ is density, $\mathbf{v}$ is velocity, P is pressure, μ is dynamic viscosity, and $\mathbf{g}$ is gravity.

15.2.4   Formulas

Torricelli's Law

Torricelli's Law relates the speed of efflux of a fluid from a small hole in a container to the height of the fluid:

$$v = \sqrt{2gh}$$

where v is the velocity, g is acceleration due to gravity, and h is the height.

Venturi Effect

The Venturi effect describes the reduction in fluid pressure when it flows through a constricted section of a pipe.

15.2.5   Working Examples

Example: Flow Rate Calculation

Calculate the flow rate of water through a pipe with given dimensions and velocity.

Example: Applications of Bernoulli's Equation

Explore practical applications of Bernoulli's equation, such as airplane lift and the functioning of a carburetor.

Chapter 16

Optoelectronics

16.1 Semiconductors and Diodes

Semiconductors and diodes play a crucial role in modern electronics, serving as key components in various electronic devices. This section explores the fundamentals of semiconductors, the behavior of diodes, and their applications.

16.1.1 Semiconductor Basics

Semiconductors are materials with conductivity between conductors and insulators. The behavior of semiconductors is described by parameters like electron mobility (μ) and intrinsic carrier concentration (n_i).

Intrinsic Carrier Concentration

The intrinsic carrier concentration in a semiconductor at thermal equilibrium is given by:

$$n_i = \sqrt{N_c \cdot N_v} \cdot e^{-\frac{E_g}{2kT}}$$

where N_c and N_v are effective density of states, E_g is the energy bandgap, k is Boltzmann's constant, and T is temperature.

16.1.2   Diode Operation

A diode is a semiconductor device with two terminals, an anode and a cathode. It operates based on the flow of charge carriers across the p-n junction.

Diode Equation

The current-voltage relationship in a diode is described by the Shockley diode equation:

$$I = I_s \left(e^{\frac{V}{nV_T}} - 1 \right)$$

where I is the diode current, I_s is the reverse saturation current, V is the voltage across the diode, n is the ideality factor, and V_T is the thermal voltage.

16.1.3   Working Examples

Example: Semiconductor Doping

Explore the impact of doping on semiconductor properties by calculating the carrier concentration for various doping levels.

Example: Diode Rectification

Investigate the role of diodes in rectification by analyzing the current-voltage characteristics of a diode in a rectifier circuit.

16.1.4   Applications

Semiconductors and diodes find applications in a variety of electronic circuits, including rectifiers, amplifiers, and light-emitting diodes (LEDs).

16.2   Transistors and Amplifiers

Transistors are fundamental semiconductor devices that form the building blocks of electronic circuits. This section explores the operation of transistors and their applications in amplifier circuits.

16.2.1 Transistor Basics

Transistors come in two main types: bipolar junction transistors (BJTs) and field-effect transistors (FETs). The behavior of a BJT is described by the following equations:

Bipolar Junction Transistor (BJT)

The BJT current-voltage relationship is given by the following equations:

$$I_C = I_S \left(e^{\frac{V_{BE}}{V_T}} - 1 \right)$$

$$I_C = \beta I_B$$

where I_C is the collector current, I_S is the saturation current, V_{BE} is the base-emitter voltage, V_T is the thermal voltage, I_B is the base current, and β is the current gain.

Field-Effect Transistor (FET)

The FET current-voltage relationship is given by the following equations:

$$I_D = I_{DSS} \left(1 - \frac{V_{GS}}{V_P} \right)^2$$

$$I_D = k(V_{GS} - V_{TH})^2$$

where I_D is the drain current, I_{DSS} is the saturation current, V_{GS} is the gate-source voltage, V_P is the pinch-off voltage, V_{TH} is the threshold voltage, and k is the transconductance parameter.

16.2.2 Amplifier Circuits

Amplifiers are circuits that increase the amplitude of signals. Common amplifier configurations include common-emitter amplifiers for BJTs and common-source amplifiers for FETs.

Common-Emitter Amplifier

The voltage gain (A_v) of a common-emitter amplifier is given by:

$$A_v = -\frac{R_C}{r_e}$$

where R_C is the collector resistor and r_e is the emitter resistance.

Common-Source Amplifier

The voltage gain (A_v) of a common-source amplifier is given by:

$$A_v = -g_m \cdot (r_d || R_D)$$

where g_m is the transconductance, r_d is the drain-source resistance, and R_D is the drain resistor.

16.2.3   Working Examples

Example: BJT Amplifier Design

Design a common-emitter amplifier for a specified voltage gain and analyze its performance.

Example: FET Amplifier Simulation

Simulate the performance of a common-source amplifier using a FET and evaluate its voltage gain.

Chapter 17

Quantum Optics

17.1 Quantum Entanglement

Quantum entanglement is a fascinating phenomenon in quantum mechanics where two or more particles become correlated in such a way that the state of one particle is directly related to the state of the others, regardless of the distance between them.

17.1.1 Entangled States

Mathematically, the entangled state of two particles is represented as:

$$|\psi\rangle = a |0\rangle |1\rangle + b |1\rangle |0\rangle$$

where $|0\rangle$ and $|1\rangle$ are the possible states of each particle, and a and b are probability amplitudes. Notably, the total probability is conserved: $|a|^2 + |b|^2 = 1$.

17.1.2   Bell States

Bell states are specific examples of entangled states that play a crucial role in quantum information processing. The four Bell states are given by:

$$\left|\Phi^+\right\rangle = \frac{1}{\sqrt{2}}(\left|0\right\rangle\left|0\right\rangle + \left|1\right\rangle\left|1\right\rangle)$$

$$\left|\Phi^-\right\rangle = \frac{1}{\sqrt{2}}(\left|0\right\rangle\left|0\right\rangle - \left|1\right\rangle\left|1\right\rangle)$$

$$\left|\Psi^+\right\rangle = \frac{1}{\sqrt{2}}(\left|0\right\rangle\left|1\right\rangle + \left|1\right\rangle\left|0\right\rangle)$$

$$\left|\Psi^-\right\rangle = \frac{1}{\sqrt{2}}(\left|0\right\rangle\left|1\right\rangle - \left|1\right\rangle\left|0\right\rangle)$$

17.1.3   Quantum Entanglement and Measurement

When a measurement is performed on one entangled particle, the state of the other particle is instantaneously determined, violating classical notions of locality. This phenomenon has been experimentally verified through various tests of Bell inequalities.

17.1.4   Working Examples

Example: Creating Entangled Photons

Describe a laboratory setup to generate entangled photons using a nonlinear crystal and verify their entanglement through coincidence measurements.

Example: Quantum Teleportation

Discuss the quantum teleportation protocol, demonstrating how the entangled state is used to transmit quantum information between distant locations.

17.2   Quantum Computing

Quantum computing is an exciting and rapidly advancing field that leverages the principles of quantum mechanics to perform computations beyond the capabilities of classical computers.

17.2.1 Qubits and Quantum Gates

In classical computing, bits can exist in one of two states: 0 or 1. Quantum computing uses quantum bits or qubits, which can exist in a superposition of both 0 and 1 simultaneously. Quantum gates manipulate qubits, and the combination of these gates allows for complex quantum computations.

17.2.2 Quantum Superposition

The principle of superposition allows qubits to exist in multiple states at once. Mathematically, a qubit $|\psi\rangle$ can be represented as:

$$|\psi\rangle = \alpha |0\rangle + \beta |1\rangle$$

where α and β are probability amplitudes.

17.2.3 Entanglement in Quantum Computing

Entanglement is a key resource in quantum computing. Entangled qubits are highly correlated, and changes to one qubit instantaneously affect the other, enabling faster and more efficient quantum algorithms.

17.2.4 Quantum Algorithms

Example: Shor's Algorithm

Shor's algorithm is a quantum algorithm that efficiently factors large numbers, posing a threat to classical cryptographic systems. It demonstrates the potential advantage of quantum computers over classical ones for certain tasks.

Example: Grover's Algorithm

Grover's algorithm provides a quadratic speedup for unstructured search problems. It can be used to search an unsorted database faster than the best classical algorithms.

17.2.5   Quantum Error Correction

Quantum computers are susceptible to errors due to decoherence and other quantum effects. Quantum error correction codes, such as the surface code, are essential for mitigating errors and ensuring the reliability of quantum computations.

17.2.6   Quantum Computing Technologies

Describe various quantum computing technologies, including superconducting qubits, trapped ions, and topological qubits, highlighting their advantages and challenges.

17.2.7   Future Prospects and Challenges

Discuss the current state of quantum computing, potential applications, and challenges, such as decoherence, noise, and scalability.

Chapter 18

Appendices

18.1 Mathematical Formulas

This section compiles essential mathematical formulas commonly used in physics. Each formula is presented along with relevant explanations, and working examples are provided for better understanding.

18.1.1 Algebraic Formulas

Quadratic Formula

The quadratic formula is used to find the roots of a quadratic equation $ax^2 + bx + c = 0$:

$$x = \frac{-b \pm \sqrt{b^2 - 4ac}}{2a}$$

Binomial Theorem

The binomial theorem expands expressions of the form $(a + b)^n$:

$$(a + b)^n = \sum_{k=0}^{n} \binom{n}{k} a^{n-k} b^k$$

18.1.2   Calculus Formulas

Derivative of a Function

The derivative of a function $f(x)$ with respect to x is denoted as $f'(x)$:

$$f'(x) = \lim_{h \to 0} \frac{f(x+h) - f(x)}{h}$$

Integral of a Function

The integral of a function $f(x)$ from a to b is denoted as $\int_a^b f(x)\,dx$:

$$\int_a^b f(x)\,dx = F(b) - F(a)$$

where $F(x)$ is the antiderivative of $f(x)$.

18.1.3   Vector Calculus Formulas

Gradient of a Scalar Field

The gradient of a scalar field V is denoted as ∇V and represents the vector of partial derivatives:

$$\nabla V = \left(\frac{\partial V}{\partial x}, \frac{\partial V}{\partial y}, \frac{\partial V}{\partial z} \right)$$

Divergence of a Vector Field

The divergence of a vector field $\mathbf{F} = (F_x, F_y, F_z)$ is denoted as $\nabla \cdot \mathbf{F}$:

$$\nabla \cdot \mathbf{F} = \frac{\partial F_x}{\partial x} + \frac{\partial F_y}{\partial y} + \frac{\partial F_z}{\partial z}$$

18.1.4   Linear Algebra Formulas

Matrix Determinant

The determinant of a 2x2 matrix $\begin{bmatrix} a & b \\ c & d \end{bmatrix}$ is given by:

$$\det = ad - bc$$

Eigenvalue and Eigenvector

For a square matrix $\mathbf{A}$, an eigenvector $\mathbf{v}$ and corresponding eigenvalue λ satisfy $\mathbf{A}\mathbf{v} = \lambda\mathbf{v}$.

18.1.5 Complex Analysis Formulas

Euler's Formula

Euler's formula relates complex exponentials to trigonometric functions:

$$e^{i\theta} = \cos(\theta) + i\sin(\theta)$$

Cauchy's Residue Theorem

Cauchy's residue theorem states that for a function analytic in a simply connected domain except at isolated singularities, the contour integral of the function is equal to $2\pi i$ times the sum of the residues at those singularities.

18.1.6 Statistics and Probability Formulas

Mean and Variance

For a set of data values x_i with corresponding frequencies f_i, the mean $\bar{x}$ and variance σ^2 are given by:

$$\bar{x} = \frac{\sum_i x_i f_i}{\sum_i f_i}$$

$$\sigma^2 = \frac{\sum_i (x_i - \bar{x})^2 f_i}{\sum_i f_i}$$

Probability Density Function (PDF)

The probability density function for a continuous random variable X is denoted as $f(x)$ and satisfies:

$$\int_{-\infty}^{\infty} f(x)\,dx = 1$$

18.1.7   Numerical Examples

Consider the quadratic equation $ax^2 + bx + c = 0$ with $a = 1, b = -3, c = 2$. Applying the quadratic formula:

$$x = \frac{-(-3) \pm \sqrt{(-3)^2 - 4(1)(2)}}{2(1)}$$

Solving, we find two roots $x_1 = 2$ and $x_2 = 1$.

For the vector field $\mathbf{F} = (2x, y, -z)$, calculate the divergence:

$$\nabla \cdot \mathbf{F} = \frac{\partial(2x)}{\partial x} + \frac{\partial y}{\partial y} + \frac{\partial(-z)}{\partial z}$$

Simplifying, we get $\nabla \cdot \mathbf{F} = 3$.

These examples illustrate the application of mathematical formulas in solving practical problems encountered in physics.

18.2   Physical Constants

This section presents a list of important physical constants frequently used in physics. These constants play a crucial role in various equations and formulas across different branches of physics.

18.2.1   Fundamental Constants

Speed of Light (c)

The speed of light in a vacuum is a fundamental constant:

$$c = 299{,}792{,}458 \, \text{m/s}$$

Planck's Constant (h)

Planck's constant relates the energy of a photon to its frequency:

$$h = 6.62607015 \times 10^{-34} \, \text{J} \cdot \text{s}$$

Gravitational Constant (G)

The gravitational constant determines the strength of the gravitational force:

$$G = 6.67430 \times 10^{-11} \, \text{m}^3/\text{kg} \cdot \text{s}^2$$

Elementary Charge (e)

The elementary charge represents the charge of a single electron:

$$e = 1.602176634 \times 10^{-19} \, \text{C}$$

18.2.2 Electromagnetic Constants

Permittivity of Free Space (ε_0):

The permittivity of free space characterizes the electric field in a vacuum:

$$\varepsilon_0 = 8.854187817 \times 10^{-12} \, \text{F/m}$$

Permeability of Free Space (μ_0):

The permeability of free space characterizes the magnetic field in a vacuum:

$$\mu_0 = 4\pi \times 10^{-7} \, \text{T} \cdot \text{m/A}$$

18.2.3 Atomic and Nuclear Constants

Avogadro's Number (N_A):

Avogadro's number represents the number of atoms or molecules in one mole:

$$N_A = 6.02214076 \times 10^{23} \, \text{mol}^{-1}$$

Boltzmann Constant (k_B):

The Boltzmann constant relates the average kinetic energy of particles in a gas to temperature:

$$k_B = 1.380649 \times 10^{-23} \, \text{J/K}$$

Rest Mass of Electron (m_e):

The rest mass of an electron is a fundamental constant:

$$m_e = 9.10938356 \times 10^{-31}\,\text{kg}$$

Rest Mass of Proton (m_p):

The rest mass of a proton is a fundamental constant:

$$m_p = 1.6726219 \times 10^{-27}\,\text{kg}$$

18.2.4   Numerical Examples

Consider a scenario where the speed of light c is utilized to calculate the energy (E) of a photon with frequency (ν):

$$E = h\nu$$

If $\nu = 5 \times 10^{14}\,\text{Hz}$, the energy is given by:

$$E = (6.62607015 \times 10^{-34}\,\text{J} \cdot \text{s}) \times (5 \times 10^{14}\,\text{Hz})$$

Another example involves Avogadro's number N_A in the context of moles and atoms. If you have 2 moles of a substance, the corresponding number of atoms is:

$$\text{Number of atoms} = 2 \times N_A$$

These examples illustrate how physical constants are applied in physics calculations.

18.3   SI Unit Conversion Tables

This section presents tables for converting between various SI units. These conversion tables are essential for physicists and engineers to navigate and apply the appropriate units in different contexts.

18.3.1   Length Conversion Table

Unit	Symbol	Conversion Factor
Meter	m	1
Kilometer	km	$1,000$
Centimeter	cm	0.01
Millimeter	mm	0.001
Micrometer	μm	1×10^{-6}
Nanometer	nm	1×10^{-9}

18.3.2   Mass Conversion Table

Unit	Symbol	Conversion Factor
Kilogram	kg	1
Gram	g	0.001
Milligram	mg	1×10^{-6}
Microgram	μg	1×10^{-9}
Tonne	t	$1,000$

18.3.3   Time Conversion Table

Unit	Symbol	Conversion Factor
Second	s	1
Millisecond	ms	1×10^{-3}
Microsecond	μs	1×10^{-6}
Nanosecond	ns	1×10^{-9}

18.3.4   Numerical Examples

Consider a scenario where a length of 2 kilometers (km) needs to be converted to meters (m):

$$\text{Length in meters} = \text{Length in kilometers} \times \text{Conversion Factor}$$

Another example involves converting a mass of 500 grams (g) to kilograms (kg):

$$\text{Mass in kilograms} = \text{Mass in grams} \times \text{Conversion Factor}$$

These examples demonstrate the application of conversion tables in practical unit conversions.

www.ingramcontent.com/pod-product-compliance
Lightning Source LLC
Chambersburg PA
CBHW080945290526

45795CB00009B/2926